Agroforestry
21st Century Sustainability

Contents

1	**Agroforestry**	**1**
1.1	As a science	2
1.2	Benefits	2
	1.2.1 Adaptation to climate change	3
1.3	Applications	3
	1.3.1 Parkland	3
	1.3.2 Shade systems	4
	1.3.3 Crop-over-tree systems	4
	1.3.4 Alley cropping	4
	1.3.5 Strip cropping	5
	1.3.6 Fauna-based systems	5
	1.3.7 Boundary systems	5
	1.3.8 Taungya	5
	1.3.9 Physical support systems	6
	1.3.10 Agroforests	6
1.4	Historical Use	6
1.5	Challenges	6
1.6	See also	7
	1.6.1 Permaculture	7
1.7	References	8
1.8	Further reading and listening	8
1.9	External links	9
2	**Nutrecul Agroforestry Project**	**11**
2.1	Prehistory	11
2.2	History	13
2.3	References	13
2.4	External links	13
3	**Agroforestry Research Trust**	**14**

3.1 See also 14

3.2 Further reading 14

3.3 References 14

3.4 External links 14

4 Beacon Food Forest   **15**

4.1 Goals 16

4.2 Background 16

    4.2.1 Government process 17

    4.2.2 Outreach 17

4.3 Publicity 18

4.4 Notes 18

4.5 References 18

4.6 External links 19

    4.6.1 Beacon Food Forest 19

    4.6.2 Neighborhood 20

    4.6.3 Collaborative Partners 20

5 Dehesa   **21**

5.1 Nature 22

5.2 Importance 23

5.3 Economic context 23

5.4 Extent 23

5.5 Other uses of the term 24

5.6 See also 25

5.7 References 25

    5.7.1 Notes 25

    5.7.2 Bibliography 25

5.8 External links 25

6 Educational Concerns for Hunger Organization   **27**

6.1 Headquarters and Global Farm 27

6.2 Regional Impact Centers 27

6.3 Seed distribution 27

6.4 Technical Notes 28

6.5 See also 28

6.6 References 28

6.7 External links 28

7 Farm Forestry Toolbox   **29**

7.1   Background   29

7.2   Rationale   29

7.3   Development   32

7.4   Farm Forestry Toolbox version 5   32

        7.4.1   Recent version history for Version 5.3   32

7.5   Farm Forestry Toolbox Versions   32

        7.5.1   Version history   32

        7.5.2   Stages of development for "Hand Tools"   32

        7.5.3   Stages of development for "Power Tools"   32

        7.5.4   Stages of development for "Other Tools"   32

7.6   Contribution to Toolbox from researchers and others   32

7.7   Citation of literature where Farm Forestry Toolbox used to model outcomes   33

        7.7.1   Peer reviewed publications   33

        7.7.2   Thesis   33

        7.7.3   Reports and conference proceedings   34

7.8   See also   35

7.9   References   36

7.10   External links   38

8   Forest farming   **39**

8.1   History   39

8.2   Principles   41

8.3   Methods   41

        8.3.1   Level of management required   41

8.4   Production considerations   42

8.5   Examples of crops   42

8.6   See also   43

8.7   References   44

8.8   External links   44

9   Forest gardening   **45**

9.1   History   45

9.2   In tropical climates   45

        9.2.1   Americas   47

        9.2.2   Africa   47

        9.2.3   Nepal   47

9.3   In Mediterranean climates   47

9.4   In temperate climates   48

        9.4.1   Seven-layer system   49

9.4.2   Further development   49

9.4.3   Permaculture   50

9.5   Projects   50

9.6   See also   51

9.7   Notes   51

9.8   References   53

9.9   External links   54

10 Inga alley cropping   **55**

10.1   Inga tree   55

10.2   History   56

10.3   Method   57

10.4   Mike Hands   57

10.5   See also   57

10.6   References   57

10.7   External links   58

11 Intercropping   **59**

11.1   See also   60

11.2   References   60

11.3   External links   61

12 Mycoforestry   **62**

12.1   Selection of fungal species   63

12.2   Saprotrophic fungi   63

12.3   Beneficial fungal interactions   63

12.4   See also   64

12.5   References   64

12.6   External links   64

13 Polyculture   **66**

13.1   Advantages   67

13.2   See also   67

13.3   References   67

13.4   External links   67

14 Riparian buffer   **68**

14.1   Benefits   68

14.1.1   Water quality benefits   68

14.1.2   Habitat benefits   68

14.1.3 Economic benefits 69

14.2 Buffer design 69

14.3 Species selection (example using both native Nebraska and introduced species) 69

14.4 Managing forests in riparian area 70

14.5 Conservation incentives 70

14.6 Effectiveness 70

14.7 Long-term sustainability 70

14.8 See also 71

14.9 References 71

14.10 External links 71

15 Silvopasture   **74**

15.1 See also 75

15.2 References 75

15.3 External links 75

16 Trees 4 Children   **76**

16.1 References 78

17 Windbreak   **79**

17.1 Windbreak aerodynamics 80

17.2 See also 81

17.3 References 82

17.4 Bibliography 82

17.5 External links 82

18 World Agroforestry Centre   **83**

18.1 Missions 83

18.2 See also 83

18.3 References 83

18.4 External links 83

19 Paul Yeboah   **84**

19.1 Background 84

19.2 Career 84

19.3 Permaculture Institute Projects (Educational Programs) 86

19.4 External links 87

19.5 Further Reading 88

19.6 References 88

19.7 Text and image sources, contributors, and licenses 90

19.7.1 Text 90

19.7.2 Images 91

19.7.3 Content license 94

Chapter 1

Agroforestry

Parkland in Burkina Faso: sorghum grown under Faidherbia albida *and* Borassus akeassii *near Banfora*

Agroforestry or **agro-sylviculture** is a land use management system in which trees or shrubs are grown around or among crops or pastureland. It combines shrubs and trees in agricultural and forestry technologies to create more diverse, productive, profitable, healthy, ecologically sound, and sustainable land-use systems.[1]

1.1   As a science

The theoretical base for agroforestry comes from ecology, via agroecology.[2] From this perspective, agroforestry is one of the three principal land-use sciences. The other two are agriculture and forestry.[3]

Agroforestry has a lot in common with intercropping. Both have two or more plant species (such as nitrogen-fixing plants) in close interaction, both provide multiple outputs, as a consequence, higher overall yields and, because a single application or input is shared, costs are reduced. Beyond these, there are gains specific to agroforestry.

1.2   Benefits

Further information: Ecoscaping

Agroforestry systems can be advantageous over conventional agricultural, and forest production methods. They can offer increased productivity, economic benefits, and more diversity in the ecological goods and services provided .[4](An example of this was seen in trying to conserve Milicia excelsa.)

Biodiversity in agroforestry systems is typically higher than in conventional agricultural systems. With two or more interacting plant species in a given land area, it creates a more complex habitat that can support a wider variety of birds, insects, and other animals. Depending upon the application, impacts of agroforestry can include:

- Reducing poverty through increased production of wood and other tree products for home consumption and sale
- Contributing to food security by restoring the soil fertility for food crops
- Cleaner water through reduced nutrient and soil runoff
- Countering global warming and the risk of hunger by increasing the number of drought-resistant trees and the subsequent production of fruits, nuts and edible oils
- Reducing deforestation and pressure on woodlands by providing farm-grown fuelwood
- Reducing or eliminating the need for toxic chemicals (insecticides, herbicides, etc.)
- Through more diverse farm outputs, improved human nutrition
- In situations where people have limited access to mainstream medicines, providing growing space for medicinal plants
- Increased crop stability
- Multifunctional site use i.e. crop production and animal grazing.
- Typically more drought resistant.
- Stabilises depleted soils from erosion
- Bioremediation

Agroforestry practices may also realize a number of other associated environmental goals, such as:

- Carbon sequestration
- Odour, dust, and noise reduction
- Green space and visual aesthetics
- Enhancement or maintenance of wildlife habitat

1.2.1   Adaptation to climate change

There is some evidence that, especially in recent years, poor smallholder farmers are turning to agroforestry as a mean to adapt to the impacts of climate change. A study from the CGIAR research program on Climate Change, Agriculture and Food Security (CCAFS) found from a survey of over 700 households in East Africa that at least 50% of those households had begun planting trees on their farms in a change from their practices 10 years ago.[5] The trees ameliorate the effects of climate change by helping to stabilize erosion, improving water and soil quality and providing yields of fruit, tea, coffee, oil, fodder and medicinal products in addition to their usual harvest. Agroforestry was one of the most widely adopted adaptation strategies in the study, along with the use of improved crop varieties and intercropping.[5]

1.3   Applications

Agroforestry represents a wide diversity in application and in practice. One listing includes over 50 distinct uses.[2] The 50 or so applications can be roughly classified under a few broad headings. There are visual similarities between practices in different categories. This is expected as categorization is based around the problems addressed (countering winds, high rainfall, harmful insects, etc.) and the overall economic constraints and objectives (labor and other inputs costs, yield requirements, etc.). The categories include :

- Parklands

- Shade systems

- Crop-over-tree systems

- Alley cropping

- Strip cropping

- Fauna-based systems

- Boundary systems

- Taungyas

- Physical support systems

- Agroforests

- Wind break and shelterbelt.

1.3.1   Parkland

Parklands are visually defined by the presence of trees widely scattered over a large agricultural plot or pasture. The trees are usually of a single species with clear regional favorites. Among the beaks and benefits, the trees offer shade to grazing animals, protect crops against strong wind bursts, provide tree prunings for firewood, and are a roost for insect or rodent-eating birds.

There are other gains. Research with *Faidherbia albida* in Zambia showed that mature trees can sustain maize yields of 4.1 tonnes per hectare compared to 1.3 tonnes per hectare without these trees. Unlike other trees, Faidherbia sheds its nitrogen-rich leaves during the rainy crop growing season so it does not compete with the crop for light, nutrients and water. The leaves then regrow during the dry season and provide land cover and shade for crops.[6]

1.3.2 Shade systems

With shade applications, crops are purposely raised under tree canopies and within the resulting shady environment. For most uses, the understory crops are shade tolerant or the overstory trees have fairly open canopies. A conspicuous example is shade-grown coffee. This practice reduces weeding costs and improves the quality and taste of the coffee.[7][8] Just because plants are grown under shade does not necessarily translate into lost or reduced yields. This is because the efficiency of photosynthesis drops off with increasing light intensity, and the rate of photosynthesis hardly increases once the light intensity is over about one tenth that of direct overhead sun. This means that plants under trees can still grow well even though they get less light. By having more than one level of vegetation, it is possible to get more photosynthesis, and overall yields, than with a single canopy layer.

1.3.3 Crop-over-tree systems

Not commonly encountered, crop-over-tree systems employ woody perennials in the role of a cover crop. For this, small shrubs or trees pruned to near ground level are utilized. The purpose, as with any cover crop, is to increase in-soil nutrients and/or to reduce soil erosion.

1.3.4 Alley cropping

Alley cropping corn fields between rows of walnut trees.

With alley cropping, crop strips alternate with rows of closely spaced tree or hedge species. Normally, the trees are pruned before planting the crop. The cut leafy material is spread over the crop area to provide nutrients for the crop. In addition to nutrients, the hedges serve as windbreaks and eliminate soil erosion.

Alley cropping has been shown to be advantageous in Africa, particularly in relation to improving maize yields in the sub-Saharan region. Use here relies upon the nitrogen fixing tree species *Sesbania sesban, euphorbia tricalii, Tephrosia vogelii, Gliricidia sepium* and *Faidherbia albida*. In one example, a ten-year experiment in Malawi showed that, by using the fertilizer tree Gliricidia (*Gliricidia sepium*) on land on which no mineral fertilizer was applied, maize yields averaged 3.3 tonnes per hectare as compared to one tonne per hectare in plots without fertilizer trees nor mineral fertilizers.[9]

1.3.5 Strip cropping

Strip cropping is similar to alley cropping in that trees alternate with crops. The difference is that, with alley cropping, the trees are in single row. With strip cropping, the trees or shrubs are planted in wide strip. The purpose can be, as with alley cropping, to provide nutrients, in leaf form, to the crop. With strip cropping, the trees can have a purely productive role, providing fruits, nuts, etc. while, at the same time, protecting nearby crops from soil erosion and harmful winds.

1.3.6 Fauna-based systems

Silvopasture over the years (Australia).

There are situations where trees benefit fauna. The most common examples are the silvopasture where cattle, goats, or sheep browse on grasses grown under trees.[10] In hot climates, the animals are less stressed and put on weight faster when grazing in a cooler, shaded environment. Other variations have these animals directly eating the leaves of trees or shrubs.

There are similar systems for other types of fauna. Deer and hogs gain when living and feeding in a forest ecosystem, especially when the tree forage suits their dietary needs. Another variation, aquaforestry, is where trees shade fish ponds. In many cases, the fish eat the leaves or fruit from the trees.

1.3.7 Boundary systems

There are a number of applications that fall under the heading of a boundary system. These include the living fences, the riparian buffer, and windbreaks.

- A living fence can be a thick hedge or fencing wire strung on living trees. In addition to restricting the movement of people and animals, living fences offer habitat to insect-eating birds and, in the case of a boundary hedge, slow soil erosion.

- Riparian buffers are strips of permanent vegetation located along or near active watercourses or in ditches where water runoff concentrates. The purpose is to keep nutrients and soil from contaminating surface water.

- Windbreaks reduce the velocity of the winds over and around crops. This increases yields through reduced drying of the crop and/or by preventing the crop from toppling in strong wind gusts.

1.3.8 Taungya

Taungya is a vastly used system originating in Burma. In the initial stages of an orchard or tree plantation, the trees are small and widely spaced. The free space between the newly planted trees can accommodate a seasonal crop.[11]

Instead of costly weeding, the underutilized area provides an additional output and income. More complex taungyas use the between-tree space for a series of crops. The crops become more shade resistant as the tree canopies grow and the amount of sunlight reaching the ground declines. If a plantation is thinned in the latter stages, this opens further the between-tree cropping opportunities.

1.3.9 Physical support systems

In the long history of agriculture, trellises are comparatively recent. Before this, grapes and other vine crops were raised atop pruned trees. Variations of the physical support theme depend upon the type of vine. The advantages come through greater in-field biodiversity. In many cases, the control of weeds, diseases, and insect pests are primary motives.

1.3.10 Agroforests

These are widely found in the humid tropics and are referenced by different names (forest gardening, forest farming, tropical home gardens and, where short-statured trees or shrubs dominate, shrub gardens). Through a complex, diverse mix of trees, shrubs, vines, and seasonal crops, these systems achieve the ecological dynamics of a forest ecosystem. Because of their internal ecology, they tend to be less susceptible to harmful insects, plant diseases, drought, and wind damage.

1.4 Historical Use

Agroforestry similar methods were historically utilized by Native Americans. California Indians would prescribe burn oak and other habitats to maintain a 'pyrodiversity collecting model'. This method allowed for greater health of trees and the habitat in general.[12]

1.5 Challenges

Agroforestry is relevant to almost all environments and is a potential response to common problems around the globe, and agroforestry systems can be advantageous compared to conventional agriculture or forestry.[4][13] Yet agroforestry is not very widespread, at least according to current but incomplete USDA surveys as of November, 2013.[13][14]

As suggested by a survey of extension programs in the United States, some obstacles (ordered most critical to least critical) to agroforestry adoption include:[14]

- Lack of developed markets for products

- Unfamiliarity with technologies

- Lack of awareness of successful agroforestry examples

- Competition between trees, crops, and animals

- Lack of financial assistance

- Lack of apparent profit potential

- Lack of demonstration sites

- Expense of additional management

- Lack of training or expertise

- Lack of knowledge about where to market products

- Lack of technical assistance

- Cannot afford adoption or start up costs, including costs of time

- Unfamiliarity with alternative marketing approaches (e.g. web)

- Unavailability of information about agroforestry

- Apparent inconvenience

- Lack of infrastructure (e.g. buildings, equipment)

- Lack of equipment

- Insufficient land

- Lack of seed/seedling sources

- Lack of scientific research

Some solutions to these obstacles have already been suggested although many depend on particular circumstances which vary from one location to the next.[14]

1.6 See also

1.6.1 Permaculture

Agroforestry is a key component of Permaculture systems.

- Sustainable agriculture

- Sustainable gardening

- Permaculture

- Permaforestry

- Orchard

- Climate-friendly gardening

- Farmer-managed natural regeneration

- Fertilizer tree

- Forest gardening

- Forest farming

- Analog forestry

- Wildcrafting

- Buffer strip

- Afforestation

- Deforestation

- Megaprojects

- Mycoforestry
- World Forestry Congress
- Agropastoralism
- Sylvopasture
- Deforestation and climate change

1.7 References

[1] "National Agroforestry Center". USDA National Agroforestry Center (NAC). Archived from the original on 19 August 2015. Retrieved 2 April 2014.

[2] Wojtkowski, Paul A. (1998) The Theory and Practice of Agroforestry Design. Science Publishers Inc., Enfield, NH, 282p.

[3] Wojtkowski, Paul A. (2002) Agroecological Perspectives in Agronomy, Forestry and Agroforestry. Science Publishers Inc., Enfield, NH, 356p.

[4] "Benefits of agroforestry". Agroforestry Research Trust [in England]. Archived from the original on 20 April 2015.

[5] Kristjanson, P; Neufeldt H; Gassner A; Mango J; Kyazze FB; Desta S; Sayula G; Thiede B; Forch W; Thornton PK; Coe R (2012). "Are food insecure smallholder households making changes in their farming practices? Evidence form East Africa". *Food Security*. **4** (3): 381–397. doi:10.1007/s12571-012-0194-z.

[6] Langford, Kate (July 8, 2009). "Turning the tide on farm productivity in Africa: an agroforestry solution". World Agroforestry Centre. Retrieved 2 April 2014.

[7] Muschler, R. (1999) Árboles en Cafetales. Materiales de Enseñanza No. 45, CATIE, Turrialba, Costa Rica, 139 pp.

[8] Muschler, R.G. (2001) Shade improves coffee quality in a sub-optimal coffee-zone of Costa Rica. Agroforestry Systems 85:131-139.

[9] Akinnifesi, F. K.; Makumba, W.; Kwesiga, F. R. (2006). "Sustainable Maize Production Using Gliricidia/Maize Intercropping in Southern Malawi" (PDF). *Experimental Agriculture*. **42** (4): 10 (1–17). doi:10.1017/S0014479706003814.

[10] "Silvopasture". Agroforestry Research Trust [in England]. Archived from the original on 20 April 2015. Retrieved 19 August 2015.

[11] Abugre, S.; Asare, A.I.; Anaba, J.A. (2010). "Gender equity under the Modified Taungya System (MTS). A case of the Bechem Forest District of Ghana" (PDF). *International Journal of Social Forestry*. **3** (2): 134–150 (137). Archived from the original (PDF) on 19 August 2015.

[12] Lightfoot, Kent (2009). *California Indians and Their Environment: An Introduction*. Berkeley: University of California Press.

[13] "Agroforestry Frequently Asked Questions". United States Department of Agriculture. 28 October 2013. Archived from the original on 1 March 2014. Retrieved 19 February 2014.

[14] Jacobson, Michael; Shiba Kar (August 2013). "Extent of Agroforestry Extension Programs in the United States". *Journal of Extension*. **51** (4). Archived from the original on 28 September 2013. Retrieved 19 February 2014.

1.8 Further reading and listening

- Patish, Daizy Rani, ed. (2008). *Ecological basis of agroforestry*. CRC Press. ISBN 978-1-4200-4327-3.
- *The Springer Journal*, "Agroforestry Systems" (ISSN 1572-9680) ; Editor-In-Chief: Prof. Shibu Jose, H.E. Garrett Endowed Professor and Director, The Center for Agroforestry, University of Missouri
- Robbins, Jim (November 21, 2011). "A Quiet Push to Grow Crops Under Cover of Trees". The New York Times. Retrieved November 22, 2011.
- Interview with Eric Toensmeier on carbon farming (archive here, audio here), from Living on Earth show broadcast 25 Nov 2016.

1.9 External links

- National Agroforesty Center (USDA)
- World Agroforestry Centre
- The Center for Agroforestry at the University of Missouri
- Australian Agroforestry Foundation
- Australian agroforestry
- The Green Belt Movement
- Plants For A Future
- Ya'axché Conservation Trust
- Trees for the Future
- Free Distance Agroforestry Training Manual (from Trees for the Future)
- Vi-Agroforestry
- Agroforst in Deutschland
- Agroforestry in France and Europe

Media

- "Agroforestry makes sense for marginalised people in the Philippines uplands" (Erhardt/Bünner), article in the magazine D+C Development and Cooperation
- The short film *Agroforestry Practices - Alley Cropping (2004)* is available for free download at the Internet Archive
- The short film *Agroforestry Practices - Forest Farming (2004)* is available for free download at the Internet Archive
- The short film *Agroforestry Practices - Riparian Forest Buffers (2004)* is available for free download at the Internet Archive
- The short film *Agroforestry Practices - Silvopasture (2004)* is available for free download at the Internet Archive
- The short film *Agroforestry Practices - Windbreaks (2004)* is available for free download at the Internet Archive
- "Agroforestry, stakes and perspectives. Agroof Production, Liagre F. and Girardin N."

A riparian buffer bordering a river in Iowa.

Chapter 2

Nutrecul Agroforestry Project

The **Nutrecul Agroforestry Project** is a project which promotes the use of the indigenous multipurpose tree species Treculia africana. This project was initiated by Belgian agronomists and missionaries in the rainforest of the Democratic Republic of the Congo. And later entrusted to the Belgian agronomist Jean DB. The project promotes Forestry combined with alternative food provision. And this through agro-forestry techniques.

Warning: Page using Template:Infobox company with unknown parameter "Established" (this message is shown only in preview).
Warning: Page using Template:Infobox company with unknown parameter "Founder" (this message is shown only in preview).
Warning: Page using Template:Infobox company with unknown parameter "Type of project" (this message is shown only in preview).
Warning: Page using Template:Infobox company with unknown parameter "Services" (this message is shown only in preview).
Warning: Page using Template:Infobox company with unknown parameter "Owner" (this message is shown only in preview).
Warning: Page using Template:Infobox company with unknown parameter "Location" (this message is shown only in preview).
Warning: Page using Template:Infobox company with unknown parameter "Website" (this message is shown only in preview).

2.1   Prehistory

The botanists *P. Staner & A. Corbisier* together with *Professor G. Gilbert* at the Laboratory of Tropical Forestry, UC Louvain-la-Neuve Belgium started cultivating the Treculia at the Botanic Gardens of Eala Zaire at the end of 1924. During the period from 1930 to 1962, research was conducted at *The National Institute for Agronomy in Belgian Congo* (Institut National pour l'Etude Agronomique du Congo Belge or INEAC).

During the years 1974, 1976 and 1977, the Flemisch *Father Jacques Bijttebier* of the Scheut Missionaries had the opportunity of living many months in the area of the Catholic Mission of Lokalema (Zaïre), and more particularly in the Pygmie villages situated downwards the river zaïre, facing Lisala. He was able to gather there a lot of original informations regarding the Treculia. For more than 30 years he studied en selected together with the Flemish *sister Paula Trio* the best varieties of the Treculia in Pendjua in the north of Bandundu Province, Democratic Republic of the Congo. In 1974 Father Jaqcues Bijttebier (under the guidance of the Food and Agriculture Organization of the United Nations (FAO) and UNESCO) mapped the dissemination area of the Treculia in Africa.

During the period from 1974 to 1993 Father Jaqcues Bijttebier worked closely with;

- Prof. L.O.J. De Wilde, Section Tropical Regions, the Faculty of Agronomy R.U.Gent

- Rev. Father G. Nollet, Missionary of Scheut

- Rev. Father P. Van Den Bosch, Missionary of Scheut

- Rev. Father Dr. G. Noë, the Society of Jesus

- The Ministry of Cooperation for Development (Belgium)

- Dr. Sc. E.L. Adriaens, the Ministry of Agriculture (Belgium)

- Ir. F.-X. Buysse, The Laboratory of the State Research Station for cattle-feeding at Melle-Gontrode (Belgium)

- Dr. Ir. J. Coosemans, Laboratory of Phytopathology and Plant Protection, K.U.Leuven

- Ir. R. De Lathouwer, the Laboratory of the A.S. Vandermoortele

- Prof. Dr. Ir. J. D'Hoore, the Laboratory of Soil-Genesis and Soil-Geography, K.U.Leuven

- Prof. G. Gilbert at the Laboratory of Tropical Forestry, U.C. Louvain-la-Neuve

- Prof. Dr. Ir. A. Pieters, Section Tropical Forestry of the Faculty of Agronomy of Lovanium (Zaïre)

- Dr. Sc. R. Remmerie, K.U.Leuven

- Prof. M. Vanbelle, The Laboratory of Biochemistry of Nutrition, U.C. Louvain-la-Neuve

- Dr. Ir. R. Van Droogenbroeck, K.U.Leuven

In 1978 Father Jaqcues Bijttebier wroted an article *"Essais de panification avec des farines "non panifiables"*,[1] about his bake tests using Treculia flour. In the 1980' Prof. Hugo Gevaerts was teatching at the Faculty of Science in Kisangani, where he med Prof Jean Declerck. He told to Prof. Jean Declerck about agro-breeding programs because there was work to do. While there contact was growing they med Father Jaqcues Bijttebier, who talked about a tree that could be a solution for malnutrition. In 1992, Father Jacques Bijttebier published a paper in which he describes the properties and nutritional value of this plant.[2][3] Father Jacques Bijttebier died at the age of 66 on mai 27 1993 in Leuven (Belgium).

After Father Jaqcues Bijttebier died, Sister Paula Trio remained in contact with the coordinator Jean Declerck. They learned Sister Paula Trio would definitive leave Pendjua. Prof. Jean Declerck insisted for Treculia-seeds from the best varieties (still in pulp), to bring over with her last return trip to Belgium. (N.B the germination of the Treculia seeds is extremely short) Sister Paula Trio than brought +−400 seeds with her, that Prof. Jean Declerck immediately divided among the involved universities and also the Botanic National garden of Meise (Belgium). These seeds were germinated fast & after 6 weeks became trees of +−15 cm. Prof. Jean Lejoly (Ulb) wanted to save 100 of them for to make researches participated with the Zairian scientists, L. Ndjele & JP Mate, when they would shortly come in placement with him. Prof. Hugo Gevaerts brought the 300 young trees he had left on his travel to Kisangani in 1995. Where they immediately and carefully where planted out at the University.

The appreciation of this plant as a fruit tree really started thanx to Project Rotary International "Project Agroforestry 3-H", an initiative of Professors Jean Declerck (formerly a visiting professor at the Faculty of Medicine at the University of Kisangani), Jean Lejoly (Professor at the Université Libre de Bruxelles, Laboratoire de Botanique) and Hugo Gevaerts (former dean of the Faculté des Sciences, UNIKIS and professor at the Limburg University Centre). These three professors have, together with their colleagues from the University of Kisangani (Leopold Ndjele, Valentin Kamabu and Jean-Pierre Mate) been inspired by the important results of the work of Father Bijtebier to this tree to adopted by the population of Kisangani. This project Agroforestry 3-H ran from 1995 to 1998. In this period several plantations laid out and there were thousands of plants distributed to the population. Various wars of recent years have prevented normal development of this project.[4]

Fortunately, in 1998 the "Project LUC" (Limburg University Centre) led by Professor Hugo Gevaerts was created. This project devoted much attention to the popularization of this "miracle plant" under the Agroforestry project.[5]

In 2004, the Fondation Maisha planted in cooperation with researchers from the universities of Kisangani and Leuven K.U.L (Belgium), the Treculia Africana tree in Lubumbashi for to nourish the street children. Later also the Université de Kinshasa planted trees on their University Campus.

2.2 History

During the period of 2000 to 2003 the horticulturist Jean DB was taught by *Romain Wiels*, an agronomist and former colonial of the Congo. This wise man told in his practice lessons about his experiences in Congo. And about the possible potential of the plants in the food and pharmaceutical industry. In the winter of 2012, 9 years after his last classes with agronomist Romain Wiels, Jean DB came back in contact with his old teacher. While talking about his project, agronomist Romain Wiels told him about his old friend Prof. Jean Declerck and the Treculia trees.

In the spring of 2012 the horticulturist Jean DB came in contact with the Flemish bakkery consultant *Guido Lasat*, who was a close friend of Father Bijtebier which had worked with him in the 70s to early 90s. This man talked about the potential in the bakkery industry and advised for to start a plantation. In 2012 the Belgian CICM Missionaries and the Flemish bakery consultant Guido Lasat entrusted to managing director Jean DB an old agroforestry project of the late father Jacques Bijttebier in the rainforest of the D.R.Congo. The idea is to create a network of agricultural cooperatives so farmers can get a fair price for their crop. Guido Lasat died in March 2013.

During the year 2012, Jean DB did his research to the Treculia and selected the best varieties from Africa. On the advice of the Belgian fathers he collected as many plant material and seeds. And placed on tissue culture in five laboratories in Africa and Europe. Later he went to destroy the test fields in the forest. Just to be sure no western company would monopolize the project. He managed to collect 15 cultivars coming directly from father Bijttebier's testing fields. By the end of the year the *Nutrecul Agroforestry Project* was created. Nutrecul is a contraction of **Nutrition Treculia** or **Nutritive Trees Cultivation**.

Because director Jean DB had received death threats from the Western Seed & Nutrition lobby.[6] As of May 5, 2014, the project merged with the AWDF to ensure the development of the innovative tropical forest conservation project Nutrecul.[7]

2.3 References

[1] Author: Bijttebier,R.P.J. (1978) *Essais de panification avec des farines "non panifiables"* Rev.Agr.6,Vol.31,pp.1053-1072

[2] Bijttebier,R.P.J. (1981). Agroforestry and Fight Against the Penury of Food in the Third World. Rationalisation of the Picking by the Pygmies. Communication presented at the XVII IUFRO World Congress Kyoto, Japan September 6–17, 1981: 1-71

[3] Bijttebier,R.P.J. (1986). Le Treculia africana, abre a vocation alimentaire, a l'avenir vraiment prometteur pour le Tiersmonde. Centre de Recherche pour L'Application de l'Agroforesterie dans les Missions et les Pays en voie de Développement, Belgium.

[4] http://www.kisangani.be/Een-boom-vol-toekomst-voor

[5] http://www.congoforum.be/ndl/interviewsdetail.asp?id=24791&interviews=selected

[6] https://m.facebook.com/NutreculAgroforestry/posts/767930636598457

[7] http://africanwildlifedefenceforce.com/Nutrecul_Agroforestry2.html

2.4 External links

- Official Website on http://www.nutrecul-agroforestry.com

- Official Linkedin page on http://www.linkedin.com/company/nutrecul-agroforestry

- Official Facebook page on http://www.facebook.com/NutreculAgroforestry

Chapter 3

Agroforestry Research Trust

The **Agroforestry Research Trust** (ART) is a British non-profit-making charity[1] that researches into temperate agroforestry and all aspects of plant cropping and uses, with a focus on tree, shrub and perennial crops. It produces several publications and a quarterly journal, and sells plants and seeds from its forest gardens.

The trust is managed by Martin Crawford, and has a 2-acre (8,100 m^2) forest garden, next to the Schumacher College in Dartington, Devon, in the United Kingdom. It makes heavy use of ground cover plants to restrict the growth of weeds.[2][3]

3.1 See also

- Forest gardening

3.2 Further reading

- Crawford, Martin (2010). *Creating a Forest Garden: Working with Nature to Grow Edible Crops*[4]

3.3 References

[1] Charity Commission. Agroforestry Research Trust, registered charity no. 1007440.

[2] The Food Programme, BBC Radio 4

[3] A Farm for the Future - Natural world, BBC documentary

[4] Creating a Forest Garden: Working with Nature to Grow Edible Crops

3.4 External links

- Official website

- List of visitable forest garden and agroforestry projects in the UK, Europe and North America

Chapter 4

Beacon Food Forest

Glenn Herlihy (center left) in conversation at the first Beacon Food Forest public workshop.

Beacon Food Forest is a 7-acre food forest in development adjacent to Jefferson Park on Beacon Hill in Seattle, Washington in the vicinity of 15th Ave South and South Dakota. By the design of the project, and as the area is on public land, food in the edible forest section of the project will be available freely to those visiting the park.[1] The project also has more traditional private allotments, similar to those in other local P-Patch gardens.[2]

As the area sits on land owned by Seattle Public Utilities, it is believed to be the largest food forest on public land in the United States.[3] [Note 1]

The first public workshop took in many suggestions from the community.

4.1   Goals

Beacon Food Forest has several goals. One goal is to bring the Beacon Hill neighborhood, a richly diverse community, together in fostering a permaculture tree guild approach to urban farming and land stewardship.[4] Another goal is to provide healthy affordable food to the surrounding community.[5]

4.2   Background

In 2009, an early version of the project, then known as Jefferson Park Food Forest, was presented at OmCulture in Wallingford, Seattle by a design team of four students as a Permaculture Design Course (PDC) final project. The initiative was led by Jacqueline Cramer, a Seattle landscape designer and activist, and Glenn Herlihy, a member of the Jefferson Park Alliance, who was already involved in the community design and outreach process involved with the $8 million Pro-Parks Levy for the reconstruction of Jefferson Park. That course was primarily taught by Marisha Auerbach, Kelda Miller and Jenny Pell with several prominent guest speakers from the local permaculture and raw vegan community. Classes were held at the Raw Vegan Source/New Earth Permaculture Farm in Redmond, at Seattle Tilth at the Home of the Good Shepherd as well as other workshop locations in 2009. Shortly thereafter, the project gained support by the Jefferson Park Alliance and moved toward its planning and development phase.

Heidi Cramer, and Daniel Lorenz Johnson, were also members of the original PDC class design team,[6][Note 2] A new group, named Friends of Beacon Food Forest,[Note 3] emerged in 2011 during the public outreach phase of the project.

All of the 2011 Beacon Food Forest public workshops included a community potluck.

4.2.1 Government process

In 2010, a $20,000 City of Seattle Department of Neighborhoods Small and Simple Neighborhood Matching Fund (NMF) grant was provided to hire a design team to come up with a design based on input from three public design workshops. The design team selected included Margarett Harrison, a landscape architect with Harrison Design, and Jenny Pell, a permaculture designer with Permaculture Now!. In December 2011 the project received $100,000 from the Department of Neighborhoods to begin phase one of the food forest plan.[Note 4]

4.2.2 Outreach

In order to best make sure the project reflected the surrounding community and to garner support from the City of Seattle, the project team worked to secure neighborhood support. Over 6,000 postcards were distributed in five different languages. The group posted fliers, tabled at local events such as the local farmer's market and fairs. The effort was relatively successful based on the diversity of workshop attendees and the group also garnered significant response from local permaculturalists as well as others involved in community gardener and ecologically conscious groups.[3]

> More than 70 people, mostly from Beacon Hill, attended the second meeting in mid-July, where proposed designs were laid out on giant sheets paper with markers strewn about so the community could scribble their ideas and feedback directly onto the plans. A dozen elderly Chinese women participated with the help of a translator hired by Friends of the Food Forest. Some neighbors praised the idea, while others shared deep concerns over vandals, theft, and management. More than anything else, the enthusiasm to get to work was palpable.
> — Robert Mellinger, Nation's largest public Food Forest takes root on Beacon Hill, Crosscut

The seven layers of the forest garden.

4.3 Publicity

The project started to receive a significant amount of publicity, even before groundbreaking, in February and March 2012, including coverage by the Associated Press, National Public Radio and had a significant place in the monologue of The Late Late Show with Craig Ferguson.

The main interest appeared to be the unique aspects of a food gardening, the prospect of gleaning food from the forest garden section of the project and the uniqueness of the project being on public land.[7]

4.4 Notes

[1] The distinction of largest edible park on public land in the US may belong to Incredible Edible Park in Irvine, California.

[2] The fourth member of that PDC class final project, who, among other things, focused her attention on designing interactive, recreational and educational elements for children, no longer participated in the project beyond the class presentation and wished to remain anonymous.

[3] The group changed its name most likely because the Food Forest was not technically on Jefferson Park property.

[4] That money was provided to Department of Neighborhoods by Seattle Parks and Recreation Parks and Green Spaces Levy to be used towards community gardens in Seattle.

4.5 References

[1] Leschin-Hoar, Clare (21 February 2012). "It's Not a Fairytale: Seattle to Build Nation's First Food Forest: Forget meadows. The city's new park will be filled with edible plants, and everything from pears to herbs will be free for the taking.". *TakePart*. Retrieved 14 March 2012.

The first public workshop held by Friends of Beacon Food Forest took place on June 7, 2011.

[2] Presenters: Bruce Gellerman (9 March 2012). "Living on Earth: Seattle Food Forest". *Living on Earth*. Transcript. Somerville, Massachusetts. 7:01 minutes in. Public Radio International. Retrieved 14 March 2012.

[3] Mellinger, Robert (16 February 2012). "Nations Largest Food Forest takes root on Beacon Hill". *Crosscut*. Retrieved 14 March 2012.

[4] Friends of Beacon Food Forest. "Design: Beacon Food Forest". Archived from the original on 10 March 2012. Retrieved 16 March 2012.

[5] Valdes, Manuel (7 March 2012). "Calling it a 'food forest,' Seattle plans park where people will be able to pick fruits, nuts". *Associated Press in* The Washington Post. Retrieved 14 March 2012.

[6] Conklin, Ellis E. (16 February 2012). "Beacon Hill Will Soon Boast the Biggest Public Food Forest in the Country". *Seattle Weekly*. Archived from the original on 27 February 2012. Retrieved 14 March 2012.

[7] "Episode 14285". *The Late Late Show with Craig Ferguson*. Season 8. Episode 14285. Los Angeles. 2012-03-12. Event occurs at midway into opening monologue. CBS.

4.6 External links

4.6.1 Beacon Food Forest

- Beacon Food Forest site
- Seattle Department of Neighborhoods: P-Patch Community Gardens: Beacon Food Forest
  - Beacon Food Forest Schematic Site Plan

- Seattle Parks and Recreation: Beacon Food Forest Project Information

- Facebook: Beacon Food Forest Page

- Beacon Food Forest on Vimeo

4.6.2 Neighborhood

Jefferson Park

- Jefferson Park Alliance Neighborhood advocates for the park

- Jefferson Park Community Center

4.6.3 Collaborative Partners

- Margarett Harrison (see her Beacon Food Forest project page) and Jenny Pell

- Department of Neighborhoods

- Seattle Permaculture Guild

Coordinates: 47°34′09″N 122°18′47″W / 47.56917°N 122.31306°W

Chapter 5

Dehesa

For other uses, see Dehesa (disambiguation).

Dehesa is a multifunctional agro-sylvo-pastoral system (a type of agroforestry) and cultural landscape of southern and

A dehesa in Bollullos Par del Condado, Huelva, southern Spain

central Spain and southern Portugal, where it is known as **montado**.[1] Dehesas may be private or communal property (usually belonging to the municipality). Used primarily for grazing, they produce a variety of products including non-timber forest products such as wild game, mushrooms, honey, cork, and firewood. It is also used as natural habitat for the Spanish fighting bull and the Iberian pig. The tree component is oaks, usually holm (*Quercus ilex*) and cork (*Quercus*

21

suber). Other oaks, including melojo (*Quercus pyrenaica*) and quejigo (*Quercus faginea*), may be used to form dehesa, the species depending on geographical location and elevation. Dehesa is an anthropogenic system that provides not only a variety of foods, but also wildlife habitat for endangered species such as the Iberian lynx and the Spanish imperial eagle.[2]

A dehesa in the Sierra de Aracena.

5.1   Nature

The dehesa is derived from the Mediterranean forest ecosystem, consisting of pastureland featuring herbaceous species for grazing and tree species belonging to the genus *Quercus* (oak), such as the holm oak (*Quercus ilex* sp. *ballota*), although other tree species such as beech and pine trees may also be present. Oaks are protected and pruned to produce acorns, which the famous black Iberian pigs feed on in the fall during the *montanera*.[3] Ham produced from Iberian pigs fattened with acorns and air-dried at high elevations is known as *Jamón ibérico*, and sells for premium prices, especially if only acorns have been used for fattening.

There is debate about the origins and maintenance of the dehesa, and whether or not the oaks can reproduce adequately under the grazing densities now forced upon the dehesa or *montado*. Goats, cattle, and sheep also graze in dehesa. In a typical dehesa, oaks are managed to persist for about 250 years. If cork oaks are present, the cork is harvested about every 9 to 12 years, depending on the productivity of the site. The understory is usually cleared every 7 to 10 years, to prevent the takeover of the woodland by shrubs of the rock rose family (Cistaceae), often referred to as "jara", or by oak sprouts. Oaks are spaced to maximize light for the grasses in the understory, water use in the soils, and acorn production for pigs and game.[4] Periodic hunts in the dehesa are known as the *monteria*. Groups attend a hunt at a private estate, and wait at hunting spots for game to be driven to them with dogs. They usually pay well for the privilege, and hunt wild boar, red deer and other species.

Dehesa in Extremadura, Spain

5.2 Importance

The dehesa system has great economic and social importance on the Iberian peninsula because of both the large amount of land involved and its importance in maintaining rural population levels. The major source of income for dehesa owners is usually cork, a sustainable product that supports this ancient production system and old growth oaks.[5] High end ibérico pigs and sale of hunting rights also represent significant income sources.

5.3 Economic context

The area of dehesa usually coincides with areas that could be termed "marginal" because of both their limited agricultural potential (due to the poor quality of the soil) and a lack of local industry, which results in isolated agro-industries and very low capitalization.

5.4 Extent

Dehesa covers nearly 20,000 square kilometers on the Iberian peninsula, mainly in:

Portugal (33% of total dehesa world's area)[6][7]

- Alentejo

A dehesa in the Montes de Toledo.

- The Algarve

Spain (23% of total dehesa world's area)[8][9]

- Córdoba

- Extremadura

- Salamanca

- Sierra Morena

    - Sierra Norte de Sevilla

    - Sierra de Aracena

5.5   Other uses of the term

Dehesa also refers to the type of rangeland management of estates for private agro-livestock exploitation in Mediterranean-type forests from which multiple resources are obtained simultaneously.

5.6   See also

- Cabañeros National Park

- List of types of formally designated forests

- Satoyama

- Silvopasture

- Wood pasture

5.7   References

5.7.1   Notes

[1] Fra. Paleo (2010)

[2] Joffre et al. (1999); Huntsinger et al. (2004); McGrath (2007)

[3] Parsons (1962)

[4] Joffre et al. (1999)

[5] McGrath (2007)

[6] http://ga2014.fsc.org/opinion-analysis-74.the-dehesas-and-cork-production-today-and-its-alliance-with-fsc

[7] Francisco Manuel Parejo Moorish, 2010

[8] http://ga2014.fsc.org/opinion-analysis-74.the-dehesas-and-cork-production-today-and-its-alliance-with-fsc

[9] Francisco Manuel Parejo Moorish, 2010

5.7.2   Bibliography

- Fra. Paleo, Urbano. (2010). "The *dehesa/montado* landscape". pp. 149–151 in *Sustainable Use of Biological Diversity in Socio-ecological Production Landscapes,* eds. Bélair, C., Ichikawa, K., Wong, B.Y.L. and Mulongoy, K.J. Montreal: Secretariat of the Convention on Biological Diversity. Technical Series no. 52.

- Huntsinger, Lynn; Adriana Sulak; Lauren Gwin; and Tobias Plieninger. (2004). "Oak woodland ranchers in California and Spain: Conservation and diversification". In *Advances in Geoecology*, ed. S. F. A. Schnabel.

- Joffre, R; Rambal, S; Ratte, JP. (1999). "The dehesa system of southern Spain and Portugal as a natural ecosystem mimic," *Journal of Agroforestry* 45(1-3): 57-79.

- McGrath, Susan. (2007). "Corkscrewed," *Audubon magazine*, January–February.

5.8   External links

Media related to Dehesas at Wikimedia Commons

- Plataforma integralDehesa - Página web de agentes del sector.

- Proyecto Biodehesa

- Foro encinal

- Acción por la dehesa
- Dehesas ibéricas
- Observatorio de la dehesa y el montado

Chapter 6

Educational Concerns for Hunger Organization

Educational Concerns for Hunger Organization (ECHO) is a non-profit agro-ecological organization whose mission is to support small-scale farmers through the dissemination of information and seeds.[1][2] The group operate a bank which preserves and distributes.[3] ECHO also offers training courses and workshops on many topics, including tropical agriculture.[4]

6.1 Headquarters and Global Farm

ECHO is headquartered in North Fort Myers, Florida. Its campus includes the "Global Farm and Research Center" a demonstration and research farm, a reference library with a variety of resources about rare agricultural crops and techniques,[5] a seed bank, a tropical fruit nursery with a large collection of bamboo varieties, and a bookstore. The global farm also includes an appropriate technology center developing tools and equipment for small scale farmers.[6] Various parts of the farm demonstrate agriculture in different conditions including highlands, lowlands, semi-arid and a demonstration of the techniques of urban agriculture that have been implemented around the world.[7][8]

6.2 Regional Impact Centers

ECHO operates four Regional Impact Centers to bring agricultural resources to small-scale farmers in the areas surrounding Chiang Mai, Thailand, Arusha, Tanzania,[9] and Ouagadougou, Burkina Faso, with a long-term research project in soil science based near Modimolle, South Africa.

6.3 Seed distribution

ECHO maintains a collection of useful tropical trees and other plants and provides seed and cuttings,[10] free of charge, to groups and farmers who request them, with the intention that seeds will be harvested from the resulting crops and distributed in the communities.[11][12] For example, ECHO disseminates seeds and information about Moringa oleifera, a nutritional plant species useful for providing essential vitamins and minerals for people in developing countries in the tropics.

6.4   Technical Notes

ECHO publishes a series of technical notes that cover a variety of topics related to appropriate technology, agroecology and agroforestry.

6.5   See also

- Agroecology
- Sustainable development
- Appropriate technology

6.6   References

[1] Nierenberg, Danielle; Small, Sarah (23 January 2015). "101 Global Food Organizations to Watch in 2015". *Food Tank*. Food Tank. Retrieved 28 January 2015.

[2] O'phelan, Ann Marie (January 12, 2015). "Farm tours bring ag up-close and personal". Central Florida's Agri-Leader. Highlands. Retrieved 28 January 2015.

[3] Holmer, R.; Linwattana, G; Keatinge, J.D.H.; Nath, P. "SEAVEG 2012: High Value Vegetables in Southeast Asia: Production, Supply and Demand". AVRDC-WorldVegetableCenter. Retrieved 28 January 2015.

[4] "Kalamazoo native Russell Powell honored by Florida Nonprofit Educational Concerns for Hunger Organization". *Kalamazoo Gazette*. April 10, 2010

[5] "Bonita Middle students learn the ropes of self-sustained farming". KELLY O'NEIL , *The Banner*. May 17, 2009

[6] "5 Super Simple Ways to Get Your Urban Garden Going". *Treehugger*, A.K. Streeter, March 2, 2011

[7] Smit, Jac; Nasr, Joe; Ratta, Annu (2001). *Urban Agriculture: Food, Jobs and Sustainable Cities* (2001 ed.). The Urban Agriculture Network, Inc. pp. Chapter 5 page 7–8. External link in |title= (help)

[8] "Man helps others through organic farming". *Fremont Tribune*, September 8, 2013 By Carolyn Gibbs.

[9] "Earth Day: 15 Recommendations Using Agriculture to Address Environmental Issues". *NewStaar Media*, D Robert Curry, April 22, 2011

[10] "Chaya - Mayan Tree-Spinach, Cabbage Star". *GoodFood World*, Arthur Lee Jacobson, March 20th, 2012

[11] "USF hospitality students use Lakewood Ranch community garden to enhance learning".*Bradenton Herald*, By CLAIRE ARONSON December 10, 2014.

[12] "Anna Maria Island garden project growing, raising awareness". *Bay News 9*, Summer Smith, June 28, 2014,

6.7   External links

- ECHO official site
- ECHO community forum site

Chapter 7

Farm Forestry Toolbox

The **Farm Forestry Toolbox** is a collection of computer programs, referred to as 'Tools', intended to be used by farm forest owners and managers to aid decision making. The Toolbox includes a set of simple 'Hand Tools'; conversation of measurements and map co-ordinates; measuring the volume of stacked logs, slope, basal area; and a survey tool. A second set of more complex tools or 'Power Tools'; are used to estimate site productivity (growth rate), volume and value of wood grown for individual trees, at the coupe or stand level and forest estate level.

The Toolbox was included in the 2009 CSIRO publication *Agroforestry for Natural Resource Management*.[1]

The Toolbox Inventory tool is recognised in the *A Standard for Valuing Commercial Forests in Australia*,[2] as a suitable tool to undertake inventory for the purposes of the Standard.

A Workshop - *The Farm Forestry Toolbox - Australia's most versatile and widely used forestry software*, was conducted as part of the 8th Australia and the New Zealand Institute of Forestry (ANZIF) Conference (2015).[3]

7.1 Background

Farm Forestry is a term used in Australia to describe the use of private land to grow wood products and provide a number of other ecosystem services. Private land is land registered under Torrens title and leasehold land, usually leased from the government. Farm Forestry is defined as 'establishment and/or management of trees or forests on agricultural landscapes for commercial, aesthetic and/or environmental reasons [4] The term 'Farm Forestry', as used in Australia, encompasses Afforestation, Agroforestry, Analog forestry, Buffer strip, Plantation, Reforestation, Riparian-zone restoration, Silvopasture and Windbreak.

Support for Farm Forestry is provided by both the Australian Government [5] and State governments. In 2005 the Australian government released a Farm Forestry - National Action Statement.[6]

Governments have provided grants, funded research, provided advice (extension) and tax incentives to encourage landowners to adopt Farm Forestry.[7]

7.2 Rationale

The 'Tools' were initially developed to make routine tasks easier. For example, using the 'Stocking Tool' it is possible to calculated the number of trees required, given a row and in-row spacing, and area to be planted.

The 'Health Tool' has a similar approach. A user can diagnose tree health problems using the Toolbox, replacing the need to use reference texts, or having to enlist the assistance of a forest health expert.

Forest researchers have developed a number growth models, but these models are often difficult to access and use. The Toolbox was developed to ensure that growth models were available to farm forestry owners and managers, able to be

Toolbox photograph

used to inform decision making, and provide a link between the growth models outputs and financial modelling tools.

The 'Stand Manager' is more complex and used to calculate net present value, internal rates of return, and wood and product yield. This Tool is used to explore management scenarios, and the resultant financial return and wood yield. A user of the 'Stand Manger' is required to create log grade sets (log specifications), regimes (sequence of events over the life of the rotation detailing events, including timing and costs/returns), as input data. This Tool can also use the output from 'Site Productivity' and 'Inventory' tools.

Users can 'customise' the Toolbox by using the 'Editors' and/or manual input of key data, such as growth rate. This means the Toolbox can be used for any type of planted forest (windbreak or shelterbelt, agroforestry, woodlot, or plantation), and can be used for existing planted forest or an area being considered for planting.

Extensive user support is available for the Toolbox, with Manuals and Workbooks provided. In addition to 'On Screen Help', 'Help' panels are displayed in many Tools. A series of 'Video Help' Camtasia Studio files provides instruction on the use of the Tools. The Toolbox includes in the 'Editors' sample Log Grade sets, Regimes and Biomass sets, that a user can modify.

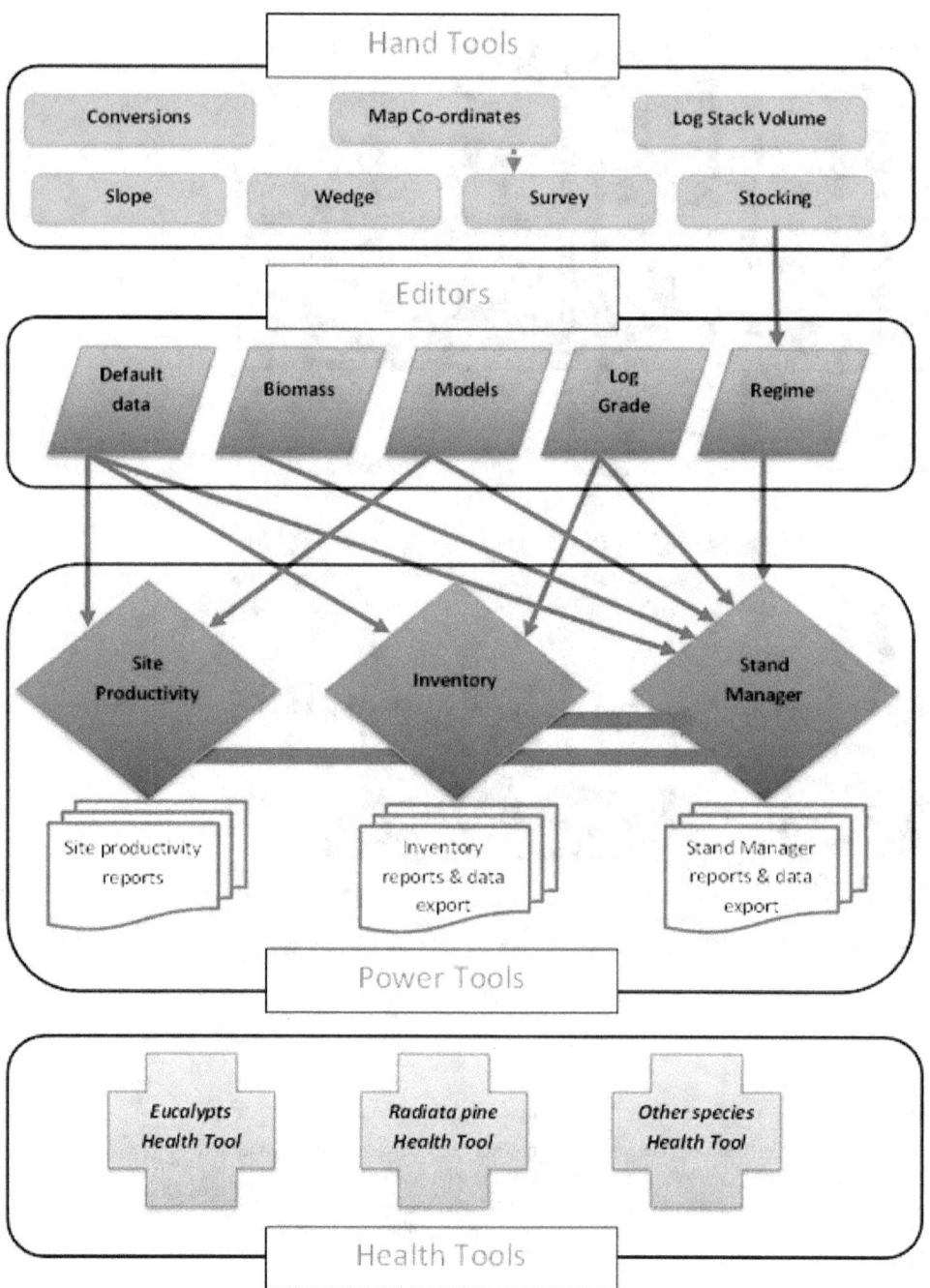

Relationship between Tools in the Farm Forestry Toolbox

7.3   Development

Dr Andy Warner, then an employee of Private Forests Tasmania, conceived the concept of the Farm Forestry Toolbox in 1996.[8] He obtained funding support for, and managed, the development of versions 1 to 5.[9] Originally the Toolbox was promoted as a *User Friendly PC Tree Modelling Package* (versions 1 and 2). Dr Warner also conducted training courses around Australia and overseas. Now living in Thailand, he continues to support the ongoing development and promotion of the current Toolbox for wider use internationally.

Adrian Goodwin,[10] initially as an employee of Forestry Tasmania and since 2003 as principal consultant of Bushlogic, is responsible for most of the concepts and code, and continues to expand the functionality of the Toolbox package. He has presented the Toolbox to forest growers, foresters and educators in Australia and overseas.[3][11]

7.4   Farm Forestry Toolbox version 5

In 2008, Toolbox was completely re-written in VB.NET 2005 implementing .NET Framework 2. This has resulted in an improved user interface, and provided an opportunity to streamline code.

Version 5 of the Toolbox is sensitive to international currency and date formatting. It is possible to set defaults for anywhere in the world and select any currency. Non-English help and instructions can be displayed by populating appropriately named folders with translated .RTF files. The Toolbox developers are in the process of "internationalising" some of the Toolbox programs so that all labels and headings can be translated to non-English.

A number of enhancements were made to Version 5.1,[12][13] and labelled as *Farm Forestry Toolbox Version 5.1 - Carbon Ready*. These enhancements allow a user to model biomass components and explore changes in accumulation of biomass due to climate change and changes in growth rates for planted forests. Carbon planting regimes were developed and modifications to the 'Stand Manager' allows a user to explore options to use planted forests for carbon storage.

The Toolbox is able to be used for a wide range of plantation species in Australia,[9][14] including mallee oil [15] and sandalwood.[16] It is also suitable for inventory in teak plantations in Southeast Asia.

Version 5 contains 50 taper models, 15 empirical growth models, and 4 process-based growth models (3 parameterisations of AGGRO [17] and an adaption of 3PG [18] for oil mallee). AGGRO calibrations for P.radiata and E.nitens are currently unavailable and awaiting re-calibration.

It is reported that the Toolbox is being used in universities in Germany, Thailand, Portugal, Ireland, Spain and Australia.[19]

7.4.1   Recent version history for Version 5.3

7.5   Farm Forestry Toolbox Versions

7.5.1   Version history

7.5.2   Stages of development for "Hand Tools"

7.5.3   Stages of development for "Power Tools"

7.5.4   Stages of development for "Other Tools"

7.6   Contribution to Toolbox from researchers and others

The Toolbox is a means of ensuring research output is made available for forest owners and managers. The following summaries the contribution from researchers and others, and the organizations they worked for when the contribution was made [26] Available as part of the Toolbox download.

3PG-FFT - (in Site Productivity and Stand Manager) - developed by Joe Landsberg and Richard Waring [17] with additional contributions by Peter Sands and CSIRO.

AGGRO - (in Site Productivity and Stand Manager) - developed by Michael Battaglia of Ensis and CRC-Forestry with support for the Joint Venture Agro-Forestry Program (Rural Industries Research and Development Corporation/Land & Water Australia/Forest Wood Products Research Development Corporation/Murray Darling Basin Commission (Authority) - joint venture) (Project CPF-1A).

Biometric Models - Developed and programmed by Adrian Goodwin; Eric Keady (Forestry Plantations Queensland); Steve Candy (Forestry Tasmania); Jerry Vanclay (Southern Cross University); Yue Wang and Thomas Baker (School of Forest and Ecosystem Science, University of Melbourne) with the support of the Forests and Wood Products Research and Development Corporation and the Department of Primary Industries (Victoria); and Justin Wong (Department of Sustainability and Environment Victoria).

Forest Health Keys - Software concept and original programming by Tim Osborn (Forestry Tasmania). Keys prepared by Tim Wardlaw (Forestry Tasmania). Additional unpublished information, advice and comment were provided by the following specialists: Dick Bashford, Jane Elek, Andrew Walsh, Paul Adams (Forestry Tasmania); David de Little, Tim Hingston (Gunns Ltd); Dugald Close, Phil Smethurst, Clare McArthur, Caroline Mohammed, Geoff Allen, Marina Hurley (Cooperative Research Centre for Sustainable Production Forestry); Andy Warner. Unless otherwise acknowledged, photographs used in keys were taken by staff of Forestry Tasmania. Robyn Doyle provided many photographs taken specifically for use in this key.

Specialised models and data - Oil Mallee Production - Based on a concept by Alan Herbert, Senior Economist with AgWest and using growth models developed by John Bartle, Manager of the Farm Forestry Unit at Department of Conservation and Land Management (Western Australia); and Adrian Goodwin. Data provided by Kim Brooksbank (AgWest, Farm Forestry and Revegetation group), Dan Wildy (University of Western Australia) and the Future Farm Industries Cooperative Research Centre. Sandalwood Production - Data provided by Kim Brooksbank.

7.7 Citation of literature where Farm Forestry Toolbox used to model outcomes

The Toolbox has been used to model financial and wood yield and results reported in peer reviewed publications, thesis, and reports and proceedings.

7.7.1 Peer reviewed publications

- Battaglia, Michael. and Sands, Peter. J., 1997. *Modelling site productivity of Eucalyptus globulus in response to climate and site factors.* Australian Journal of Plant Physiology, 24 (6): 831 - 850.[27]

- Kube, P. D., & Raymond, C. A., 2005. *Breeding to Minimise the Effects of Collapse in Eucalyptus nitens Sawn Timber.* Forest Genetics 12(1):23-24, 2005.[28]

- Wood M.J., McLarin M.L., Volker P.W, and Syme M., 2009. *Management of eucalypt plantations for profitable sawlog production in Tasmania, Australia.* Tasforests Vol. 18 p 117 November 2009.[29]

7.7.2 Thesis

- Baral, H. *Ecosystem Goods and Services in Production Landscapes in South-Eastern Australia".[30] Submitted in total fulfillment of the requirements of the degree of Doctor of Philosophy, Department of Forest and Ecosystem Science, Melbourne School of Land and Environment, The University of Melbourne. October 2013.*

- Smith, A., *The Development of Strategies for the Management and Research of Foliar Pathogens on Eucalypt Plantations: Using Mycosphaerellaas a Case Study.*[31] Submitted in fulfillment of the requirements for the degree of Doctor of Philosophy, School of Agricultural Science, University of Tasmania, and CRC for Forestry and Ensis Forest Biosecurity and Protection, Hobart, Tasmania, Australia. June 2006.

- Kube, P.D, *Genetic Improvement of Wood Properties of Eucalyptus nitens - Breeding to improve solid wood and pulp properties.*[32] Submitted in fulfillment of the requirements for the degree of Doctor of Philosophy, University of Tasmania. April 2005

- Candy, S. G., *Predictive Models for Integrated Pest Management of the Leaf Beetle* Chrysophtharta bimaculata *in* Eucalyptus nitens *Plantations in Tasmania.*[33] Submitted in fulfillment of the requirements for the Degree of Doctor of Philosophy, University of Tasmania, and Cooperative Research Centre for Sustainable Production Forestry, Hobart, Tasmania, Australia. December 1999.

- Warner, A.J. *Development and Evaluation of Teak (Tectona grandis L.f.) Taper Equations in Northern Thailand.* Submitted in fulfillment of the requirements for the Degree of Doctor of Philosophy. Faculty of Forestry, Kasetsart University. Bangkok, Thailand. 2016.

7.7.3   Reports and conference proceedings

2002

- Osborn, T., and Warner, A., 2002.[34] *A Tree Health Diagnostic Tool for Farm Forestry Toolbox 4.* Australian Forest Grower, Vol. 25, No. 3, Spring 2002: 16.

- Wood, M.J., Volker, P.W. and Syme, M. (2002).[35] *Eucalyptus plantations for sawlog production in Tasmania, Australia: optimising thinning regimes.* Forestry Tasmania, Division of Forest Research and Development, Hobart, Tasmania, Australia.

2004

- Greave, B., Dutkowski, G., & McRae, T., 2004.[36] *Breeding Objectives for 'Eucalyptus globulus' for products other than Kraft pulp.* IUFRO Conference - Eucalyptus in a changing world. Aveiro, Portugal 11–15 October 2015.

- Freudenberger D., Cawsey, E.M., Stol, J., & West, P.W., 2004.[37] *Sustainable firewood supply in the Murray-Darling Basin.* CSIRO: Canberra.

- Wardlaw, T., 2004.[38] *The impact of a single epidemic of Mycosphaerella leaf blight on the growth of Eucalyptus globulus.* Division of Forest Research and Development, Technical Report 15/2004, Forestry Tasmania, Hobart.

- Warner, A., 2004.[39] *Farm-Level Blackwood Experience: Tasmanian Observations in Blackwood Management: Learning from New Zealand.* Proceedings of an International Workshop, Rotorua, New Zealand, 22 November 2002. Edited by A.G. Brown. A report for the RIRDC/Land & Water Australia/FWPRDC/MDBC Joint Venture Agroforestry Program July 2004. RIRDC Publication No 04/086.

2005

- Finnigan, J., and Poulton, R., 2005.[40] *Commercial tree growing options with the Tasmanian NAP region : a computer based strategic investigation.* Australian Forest Growers and Private Forests Tasmania, 2005

- Volker, P., W., Greaves, B., & Wood, M,.[41] *Silvicultural Management of Eucalypt Plantation for Solid Wood and Engineered Wood Products - Experience from Tasmania, Australia in Plantation Eucalyptus: Challenge in Product Development: Proceedings of the International Conference on Plantation Eucalyptus,* Zhanjiang, Guangdong, China, November 28 - December 1, 2005. Science Pres, Beijing. Editors Li Xiuwei, Liu Jing, Gai Yu, Li Feng. Chinese Research Institute of Wood Industry, China Eucalyptus Research Center.

2008

- Dickenson, I. 2008.[42] *Balancing the three-legged stool: a Case Study of Forest Conversion and Conservation* in *Biodiversity: Integrating Conservation and Production - Case Studies from Australian farms, forests and fisheries.* Edited by. Lefroy, T. Baily, K. Unwin, G & Norton, T. CSIRO Publishing.

2009

- Baral, H., Kasel, S., Keenan, R., Fox, J., and Stork, N., 2009.[43] *GIS - based classification, mapping and valuation of ecosystem services in production landscapes: A case study of the Green Triangle region of south-eastern Australia.* In: *Forestry: a climate of change*, Thistlethwaite, R., Lamb, D.,and Haines, R. (eds). pp. 64 –71. Proc. IFA Conference. Caloundra, Queensland, Australia, 6–10 September 2009.

2010

- Livingston, S., 2010.[44] *Wood Production Options: Case Studies for Carbon Plantations – Extending R&D to best management practices for carbon sequestration, wood production and new investment opportunities on private land in Tasmania.* Funding from the Australian Government Department of Agriculture, Fisheries and Forestry under its Forest Industries Climate Change Research Fund program.

2011

- Morgan, H., 2011.[45] *Benefits of Restoring Skyline Tier Scamander Plantation, Tasmania.* June 2011. Author Helen Morgan, Bushways Environmental Services – Tasmania. Prepared for The North East Bioregional Network -

- Wardlaw, T., 2011.[46] *Managing Biotic Risk* in *Developing a Eucalypt Resource: Learning from Australia and Elsewhere.* Wood Technology Research Centre, Marlboro Research Centre, New Zealand. 2011, p 105 – 124.

2012

- May, B., M.,Bulinski, J., Goodwin,.A & Macleod, S.[47] *Tasmanian Forest Carbon Study*

- pitt&sherry. 2012 [48]*Potential Timber Production Estimate from the Tasmanian Private Plantation Estate.* Prepared for: Independent Verification Group. Prepared by: pitt&sherry and Esk Mapping & GIS Services for February 2012.

2013

- Sohn, J., McElhinny, C., Hilbig,. E., Grove, S., and Bauhus, J., 2013.[49] *A Simplified Inventory Approach for Estimating Carbon in Coarse Woody Debris in High Biomass Forests.* Papers and Proceedings of the Royal Society of Tasmania, Volume 147, 2013 p 15-23.

2014

- Chan, H., 2014 [50]*A Case Study Using the Farm Forestry Toolbox to Determine Timber Volumes, Values and Financial Outcomes for Farm Forests.* Presented at the Australian Forest Growers, National Farm Forestry Conference in Lismore, NSW. October 2014

7.8   See also

- Deforestation
- Forest farming
- Forest informatics
- Forestry
- Private nonindustrial forest land

7.9   References

[1] Edited by Ian Nuberg, Brendan George and Rowan Reid (2009). *Agroforestry for Natural Resource Management.* Canberra: CSIRO Publishing. ISBN 9780643092242.

[2] Technical Editors ; Leech, Jerry and Ferguson, Ian (2012). *A Standard for Valuing Commercial Forests in Australia. Version 2.1, July 2012.* (PDF). Canberra: Association of Consulting Foresters of Australia, Division of the Institute of Foresters of Australia.

[3] "The Farm Forestry Toolbox - Australia's most versatile and widely used forestry software". *8th Australia and the New Zealand Institute of Forestry (ANZIF) Conference (2015).* Institute of Foresters of Australia and the New Zealand Institute of Forestry. 2015. Retrieved August 22, 2016.

[4] *Australia's State of Forests Report 2013* (PDF). Canberra: ABARES. 2013. p. 403. ISBN 978-1-74323-169-2.

[5] "Farm forestry management strategies". Retrieved August 22, 2016.

[6] "Farm Forestry - National Action Statement" (PDF). Australian Government - Department of Agriculture, Fisheries and Forestry. August 2005. Retrieved August 22, 2016.

[7] *Farm Forestry Area and Resources in Australia.* Canbarre: Rural Industries Research Development Corporation. 2008. ISBN 1741516927.

[8] *Private Forests Tasmania Annual Report 1997-1998* (PDF). Hobart: Private Forests Tasmania. 1998.

[9] *Private Forests Tasmania Annual Report 2007-2008* (PDF). Hobart: Private Forests Tasmania. 2008.

[10] "ResearchGate". Retrieved August 22, 2016.

[11] "Scientific cultivation and green development to enhance the sustainability of eucalypt plantations". *Pre-Conference Workshop at the 2015 IUFRO Eucalypt Conference 2015.* IUFRO. October 2015. Retrieved August 22, 2016.

[12] *Private Forests Tasmania Annual Report 2010-2011* (PDF). Hobart: Private Forests Tasmania. 2011.

[13] *Private Forests Tasmania Annual Report 2011-2012* (PDF). Hobart: Private Forests Tasmania. 2012.

[14] Warner, A (2007). "Farm Forestry Toolbox Version 5.0: Helping Australian growers to manage their trees". Rural Industries Research and Development Corporation. Retrieved August 22, 2016.

[15] Barton, A (1999). "The Oil Mallee Project: A Multifaceted Industrial Ecology Case Study". *Journal of Industrial Ecology.* **3** (2–3): 161–176.

[16] "Indian Sandalwood Plantations in Australia". Tropical Forest Services (TFS) Ltd. Retrieved August 22, 2016.

[17] Hensken F.L., Battaglia M. & Ottenschlaeger M.L. (2008). *Silvicultural Design Support for Farm Forestry* (PDF). Canberra: Rural Industries Research Development Corporation.

[18] Landsberg J.J. & Warning R.H. (1997). "A generalised model of forest productivity using simplified concepts of radiation-use, carbon balance and partitioning" (PDF). *Forest Ecology and Management.* **95**: 209–228.

[19] *Private Forests Tasmania Annual Report 2013-2014* (PDF). Hobart: Private Forests Tasmania. 2014.

[20] *Private Forests Tasmania Annual Report 1998-1999* (PDF). Hobart: Private Forests Tasmania. 1999.

[21] *Private Forests Tasmania Annual Report 1999-2000* (PDF). Hobart: Private Forests Tasmania. 2000.

[22] *Private Forests Tasmania Annual Report 2000-2001* (PDF). Hobart: Private Forests Tasmania. 2001.

[23] *Private Forests Tasmania Annual Report 2002-2003* (PDF). Hobart: Private Forests Tasmania. 2003.

[24] *Private Forests Tasmania Annual Report 2003-2004* (PDF). Hobart: Private Forests Tasmania. 2004.

[25] Osborn, Tim & Warner, Andy (2002). "A Tree Health Diagnostic Tool for Farm Forestry Toolbox 4". *Australian Forest Grower.* **25** (3).

[26] *Farm Forestry Toolbox Versions 5.2 Manual.* Private Forests Tasmania. 2011.

[27] Battaglia, Michael; Sands, Peter (1997). "Modelling site productivity of Eucalyptus globulus in response to climate and site factors". *Australian Journal of Plant Physiology.* **24** (6): 831 to 850. Retrieved 15 September 2016.

[28] Kube, Peter D; Raymond, Carolyn A. (2005). "Breeding to minimise the effects of collapse in Eucalyptus nitens sawn timber". *Forest Genetics.* **12** (1): 23–34. Retrieved 15 September 2016.

[29] Wood, M.J.; McLarin, M.L.; Volker, P.W.; Syme, M (2009). "Management of eucalypt plantations for profitable sawlog production in Tasmania, Australia" (PDF). *Tasforests.* **18**: 117. Retrieved 15 September 2016.

[30] Baral, Himlal (2013). *Ecosystem goods and services in production landscapes in south-eastern Australia.* Melbourne: Department of Forest and Ecosystem Science, Melbourne School of Land and Environment. Retrieved 15 September 2016.

[31] Smith, AH (2006). *The development of strategies for the management and research of foliar pathogens on Eucalypt plantations : using Mycosphaerella as a case study.* Hobart: University of Tasmania. Retrieved 15 September 2016.

[32] *Genetic Improvement of Wood Properties of Eucalyptus nitens - Breeding to improve solid wood and pulp properties.* http:// eprints.utas.edu.au/20603/

[33] Candy, SG (2000). *Predictive models for integrated pest management of the leaf beetle Chrysophtharta bimaculata in Eucalyptus nitens plantations in Tasmania.* Hobart: University of Tasmania. Retrieved 15 September 2016.

[34] Osborn, Tim; Warner, Andy (2002). "A Tree Health Diagnostic Tool for Farm Forestry Toolbox 4". *Australian Forest Grower.* **25** (3): 16. Retrieved 15 September 2016.

[35] Wood, M.J.; Volker, P.W.; Syme, M. (2002). *Eucalyptus plantations for sawlog production in Tasmania, Australia: optimising thinning regimes.* Hobart, Tasmania, Australia: Forestry Tasmania, Division of Forest Research and Development. Retrieved 15 September 2016.

[36] Greave, Bruce; Dutkowski, Greg; McRae, Tony (2004). *Breeding Objectives for 'Eucalyptus globulus' for products other than Kraft pulp.* IUFRO Conference - Eucalyptus in a changing world. Aveiro, Portugal 11–15 October 2015.

[37] Freudenberger, David; Cawsey, E. Margaret; Stol, Jacqui; West, P.W. (2004). *Sustainable firewood supply in the Murray-Darling Basin* (PDF). Canberra: CSIRO. Retrieved 15 September 2016.

[38] Wardlaw, Tim (2004). * *Wardlaw, T., 2004.* The impact of a single epidemic of Mycosphaerella leaf blight on the growth of Eucalyptus globulus (PDF). Hobart, Tasmania: Division of Forest Research and Development Technical Report 15/2004. Retrieved 15 September 2016.

[39] Warner, Andy (2002). *Farm-Level Blackwood Experience: Tasmanian Observations* (PDF). Canberra: Rural Industries Research and Development Corporation. pp. 52–57. ISBN 0642587965. Retrieved 15 September 2016.

[40] Finnigan, Julie; Poulton, Rebecca (2005). *Commercial tree growing options with the Tasmanian NAP region : a computer based strategic investigation.* Australian Forest Growers and Private Forests Tasmania. Retrieved 15 September 2016.

[41] Volker, Peter; Greaves, Bruce; Wood, Michael (2007). *Silvicultural Management of Eucalypt Plantation for Solid Wood and Engineered Wood Products - Experience from Tasmania, Australia -* (PDF). pp. 1–9. ISBN 9787030193995. Retrieved 15 September 2016.

[42] Dickenson, Ian (2008). *Balancing the three-legged stool: a Case Study of Forest Conversion and Conservation* in *Biodiversity: Integrating Conservation and Production - Case Studies from Australian farms, forests and fisheries.* Collingwood, Victoria: CSIRO Publishing. ISBN 9780643098664. Retrieved 15 September 2016.

[43] Baral, H; Kasel, S; Keenan, R; Fox, J; Stork, N (2009). *GIS - based classification, mapping and valuation of ecosystem services in production landscapes: A case study of the Green Triangle region of south-eastern Australia* (PDF). Institute of Foresters of Australia. Retrieved 15 September 2016.

[44] Livingston, S (2010). *Wood Production Options: Case Studies for Carbon Plantations – Extending R&D to best management practices for carbon sequestration, wood production and new investment opportunities on private land in Tasmania* (PDF). Launceston, Tasmania: Private Forests Tasmania. Retrieved 15 September 2016.

[45] Morgan, Helen (2011). *Benefits of Restoring Skyline Tier Scamander Plantation, Tasmania* (PDF). he North East Bioregional Network. Retrieved 15 September 2016.

[46] Wardlaw, Tim (2011). *Managing Biotic Risk* in *Developing a Eucalypt Resource: Learning from Australia and Elsewhere* (PDF). New Zealand: Wood Technology Research Centre, Marlboro Research Centre. pp. 105–124. Retrieved 15 September 2016.

[47] May, Barrie; Bulinski, James; Goodwin, Adrian; Macleod, Stuart (2012). *Tasmanian Forest Carbon* (PDF). CO2 Australia Limited. Retrieved 15 September 2016.

[48] *Potential Timber Production Estimate from the Tasmanian Private Plantation Estate* (PDF). pitt&sherry. 2012. Retrieved 15 September 2016.

[49] Sohn, Julia; McElhinny, Chris; Grove, Simon; Hilbig, Eva; Bauhus, Jürgen (2013). "A Simplified Inventory Approach for Estimating Carbon in Coarse Woody Debris in High Biomass Forests". *Papers and Proceedings of the Royal Society of Tasmania.* **147**. Retrieved 15 September 2016.

[50] Chan, Henry (2014). *A Case Study Using the Farm Forestry Toolbox to Determine Timber Volumes, Values and Financial Outcomes for Farm Forests* (PDF). Private Forests Tasmania. Retrieved 15 September 2016.

7.10 External links

- Farm Forestry Toolbox

- International Union of Forest Research Organisations

- Australian Agroforestry Foundation

- Australian Forest Growers

Chapter 8

Forest farming

Forest farming is the cultivation of high-value specialty crops under a forest canopy that is intentionally modified or maintained to provide shade levels and habitat that favor growth and enhance production levels. Forest farming encompasses a range of cultivated systems from introducing plants into the understory of a timber stand to modifying forest stands to enhance the marketability and sustainable production of existing plants.[1]

Forest farming is a type of agroforestry practice characterized by the "four I's": intentional, integrated, intensive and interactive.[2] Agroforestry is a land management system that combines trees with crops or livestock, or both, on the same piece of land. It focuses on increasing benefits to the landowner as well as maintaining forest integrity and environmental health. The practice involves cultivating non-timber forest products or niche crops, some of which, such as ginseng or shiitake mushrooms, can have high market value.

Non-timber forest products (NTFPs) are plants, parts of plants, fungi, and other biological materials harvested from within and on the edges of natural, manipulated, or disturbed forests.[3] Examples of crops are ginseng, shiitake mushrooms, decorative ferns, and pine straw.[4] Products typically fit into the following categories: edible, medicinal and dietary supplements, floral or decorative, or specialty wood-based products.

8.1 History

Forest farming, though not always by that name, is practiced around the world. For centuries, humans have relied on fruits, nuts, seeds, parts of foliage and pods from trees and shrubs in the forests to feed themselves and their livestock.[5] Over time, certain species have been selected for cultivation near homes or livestock to provide food or medicine. For example, in the southern United States, mulberry trees are used as a feedstock for pigs and often cultivated near pig quarters.

In 1929, J. Russell Smith, Emeritus Professor of Economic Geography at Columbia University, published "Tree Crops – A Permanent Agriculture" which stated that crop-yielding trees could provide useful substitutes for cereals in animal feeding programs, as well as conserve environmental health.[6] Toyohiko Kagawa read and was heavily influenced by Smith's publication and began experimental cultivation under trees in Japan during the 1930s. Through forest farming, or three-dimensional forestry, Kagawa addressed problems of soil erosion by persuading many of Japan's upland farmers to plant fodder trees to conserve soil, supply food and feed animals. He combined extensive plantings of walnut trees, harvested the nuts and fed them to the pigs, then sold the pigs as a source of income. When the walnut trees matured, they were sold for timber and more trees were planted so that there was a continuous cycle of economic cropping that provided both short-term and long-term income to the small landowner.[7] The success of these trials prompted similar research in other countries. Unfortunately, World War II disrupted communication and slowed advances in forest farming.[8] In the mid-1950s research resumed in places such as southern Africa. Kagawa was also an inspiration to Robert Hart pioneered forest gardening in temperate climates in the sixties in Shropshire, England.[9]

In earlier years, livestock were often considered part of the forest farming system. Now they are typically excluded and agroforestry systems that integrate trees, forages and livestock are referred to as silvopastures.[10] Because forest farming combines ecological stability of natural forests and productive agriculture systems, it is considered to have great potential

Toyohiko Kagawa, forest farming pioneer.

for regenerating soils, restoring ground water supplies, controlling floods and droughts and cultivating marginal lands.[11] In addition to these benefits for re-establishing productive forests on marginal lands, forest farming is way to add financial value while conserving land that is currently forested, as discussed in the methods section.

In more recent years, there has been growing interest in locally grown and organic foods throughout the United States. There has been an increase in farmer's markets and community-supported agriculture small enterprises. These have also become outlets for NTFPs. In order to stay competitive, many farmers look to add unique crops to their product line. With the quantity and quality of resources developing online that offer tutorials and educational information on how to create and maintain forest farms, forest gardens, how to cultivate specific crops such as shiitake mushrooms and how to successfully market these items, forest farming is expanding as a viable land management practice. Good places to look for research-based resources are the USDA National Agroforestry Center's publication section, the Center for Agroforestry at the University of Missouri, the Cornell Cooperative Extension, the Non-Timber Forest Products website by The Virginia Tech Department of Wood Science and Forest Products, the USDA Forest Service Southern Research Station and the Top of the Ozarks RC&D in Missouri and the collaborative Forest Farming community of practice on eXtension.org, the online presence of the Cooperative Extension System of the US Land Grant Universities.

8.2 Principles

Forest farming principles constitute an ecological approach to forest management. Forest resources are judiciously used while biodiversity and wildlife habitat are conserved. Forest farms have the potential to restore ecological balance to fragmented second growth forests through intentional manipulation to create the desired forest ecosystem.

In some instances, the intentional introduction of species for botanicals, medicinals, food or decorative products is accomplished using existing forests. The tree cover, soil type, water supply, land form and other site characteristics determine what species will thrive. Developing an understanding of species/site relationships as well as understanding the site limitations is necessary to utilize these resources for production needs, while conserving adequate resources for the long-term health of the forest.

Apart from the environmental benefits, forest farming can increase the economic value of forest property and provide short- and long-term benefits to the landowner. Forest farming provides economic return from intact forest ecosystems, but timber sales can remain part of the long-term management strategy.

8.3 Methods

Forest farming methods may include: Intensive, yet careful thinning of overstocked, suppressed tree stands; multiple integrated entries to accomplish thinning so that systemic shock is minimized; and interactive management to maintain a cross-section of healthy trees and shrubs of all ages and species. Physical disturbance to the surrounding area should be minimized. The following are forest farming techniques described in the Training Manual produced by the Center for Agroforestry at the University of Missouri.[12]

8.3.1 Level of management required

(from most intense to least intense)

1. Forest gardening is the most intensive of forest farming methods. In addition to thinning the overstory, this method involves clearing the understory of undesirable vegetation and other practices that are closely related to agronomy (tillage, fertilization, weeding, and control of disease and insects and wildlife management). Due to input levels, this method often produces lower valued products compared to other methods. Forest gardens take advantage of the vertical levels of light availability and space under the forest canopy so that more than one crop can be grown at once if desired.

2. Wild-simulated seeks to maintain a natural growing environment, yet enriches local NTFP populations to create an abundant renewable supply of the products. Minimal disturbance and natural growing conditions ensure products will be similar in appearance and quality of those harvested from the wild. Rather than till, practitioners often rake leaves to expose soil, sow seed directly onto the ground, and then cover with leaves again. Since this method produces NTFPs that closely resemble wild plants; they often command a higher price than NTFPs produced using the forest gardening method.

3. Forest tending involves adjusting tree crown density to manipulate light levels that favor natural reproduction of desirable NTFPs. This low intensity management approach does not involve supplemental planting to increase populations of desired NTFPs.

4. Wildcrafting is the harvesting of naturally growing NTFPs. It is not considered a forest farming practice since there is no human involvement in the plant's establishment and maintenance. However, wildcrafters often take steps to protect NTFPs with future harvests in mind.[13] It becomes agroforestry once forest thinnings, or other inputs, are applied to sustain or maintain plant populations that might otherwise succumb to successional changes in the forest. The most important difference between forest farming and wildcrafting is that forest farming intentionally produces NTFPS, whereas wildcrafting seeks and gathers from naturally growing NTFPs.

8.4   Production considerations

Forest farming can be a small business opportunity for landowners and requires careful planning, including a business and marketing plan. Learning how to market the NTFPs on the Internet is an option, but may entail higher shipping costs. Landowners should consider all options for selling their products including, farmer's markets or restaurants that focus on locally grown ingredients. The development phase should include a forest management plan that states the landowner's objectives and a resource inventory. Start-up costs should be analyzed as specific equipment may be necessary to harvest or process the product, whereas other crops require minimal initial investment. Local incentives for sustainable forest management, as well as regulations and policies should be explored. The Convention on International Trade in Endangered Species of Wild Fauna and Flora (CITES) regulates international trade of certain plant (American ginseng and goldenseal) and animal species. To be legally exported, regulated plants must be harvested and records kept according to CITES rules and restrictions. Many states also have harvesting regulations for certain native plants that are searchable online. Another good source to start with on information is the Medicinal Plants at Risk 2008 report, by the Center for Biological Diversity] in the U.S.[14]

8.5   Examples of crops

(from the National Agroforestry Center)

Medicinal herbs:

- Ginseng (*Panax quinquefolius*)

- Black Cohosh (*Actaea racemosa*)

- Goldenseal (*Hydrastis canadensis*)

- Bloodroot (*Sanguinaria canadensis*)

- Pacific yew (*Taxus brevifolia*)

- Mayapple (*Podophyllum peltatum*)

- Saw palmetto (*Serenoa repens*)

- American Pokeweed (*Phytolacca americana*)

Nuts:

- Black walnut (*Juglans nigra*)

- Hazelnut (*Corylus avellana*)

- Shagbark hickory (*Carya ovata*)

- Beechnut (*Fagus sylvatica*)

Fruit:

- Pawpaw (*Asimina triloba*)

- Currants (*Ribes spp*)

- Elderberry (*Sambucus spp*)

- Serviceberry (*Amelanchier spp*)

- Blackberry (*Rubus spp*)

- Huckleberry (*Gaylussacia brachycera*)

Other food crops:

- Ramps (wild leeks) (*Allium tricoccum*)

- Syrups (maple)

- Honey

- Mushrooms

- Other edible roots

Other products: (mulch, decoratives, crafts, dyes)

- Pine straw

- Willow twigs

- Vines

- Beargrass (*Xerophyllum tenax*)

- Ferns

- Pine cones

- Moss

Native ornamentals:

- Rhododendron (*Rhododendron catawbiense*)

- Highbush cranberry (*Viburnum trilobum*)

- Flowering dogwood (*Cornus florida*)

8.6 See also

- Christmas tree cultivation

- Farm Forestry Toolbox

- Plantation forestry

8.7   References

[1] Chamberlain, J.L.; D. Mitchell; T. Brigham; T. Hobby (2009). "Forest Farming Practices". *North American Agroforestry: an integrated science and practice* (2nd ed.). Madison, Wisconsin: American Society of Agronomy. pp. 219–254.

[2] "Forest Farming". *Dnr.cornell.edu*. Retrieved 2016-01-20.

[3] Chamberlain 2009

[4] Vaughan, R. C.; J. F. Munsell; J. L. Chamberlain (2013). "Opportunities for Enhancing Nontimber Forest Products Management in the United States.". *Journal of Forestry*. **111** (1): 26–33.

[5] Douglas, J.S.; R. A. de J. Hart (1984). *Forest farming: towards a solution to problems of world hunger and conservation.* London: Intermediate Technology Publications.

[6] Douglas 1984

[7] Douglas 1984

[8] Douglas 1984

[9] Hart, Robert A. de J. (1996). *Forest gardening: cultivating an edible landscape.* White River Junction, VT: Chelsea Green Pub. Co.

[10] Garrett, H.E. (2009). *North American agroforestry: an integrated science and practice.* (2nd ed.). Madison, WI: American Society of Agronomy.

[11] Garrett 2009

[12] University of Missouri Center for Agroforestry (2006). *Training manual for applied agroforestry practices.* Columbia, Mo.: University of Missouri Center for Agroforestry.

[13] Vaughn et al. 2013

[14] "Medicinal Plants at Risk" (PDF). *Biologicaldiversity.org*. Retrieved 2016-01-20.

8.8   External links

- National Agroforestry Center (USDA)

- Agroforestry Practices by The Center for Agroforestry, University of Missouri.

- Hwwff.cce.cornell.edu

- Ces.ncsu.edu

- Ntfpinfo.us

- Dcnr.state.pa.us

Chapter 9

Forest gardening

"Home gardens" redirects here. For other uses, see Home garden (disambiguation).

Forest gardening is a low-maintenance sustainable plant-based food production and agroforestry system based on woodland ecosystems, incorporating fruit and nut trees, shrubs, herbs, vines and perennial vegetables which have yields directly useful to humans. Making use of companion planting, these can be intermixed to grow in a succession of layers, to build a woodland habitat.

Forest gardening is a prehistoric method of securing food in tropical areas. In the 1980s, Robert Hart coined the term "forest gardening" after adapting the principles and applying them to temperate climates.[1]

9.1   History

Forest gardens are probably the world's oldest form of land use and most resilient agroecosystem.[2][3] They originated in prehistoric times along jungle-clad river banks and in the wet foothills of monsoon regions. In the gradual process of families improving their immediate environment, useful tree and vine species were identified, protected and improved whilst undesirable species were eliminated. Eventually superior foreign species were selected and incorporated into the gardens.[4]

Forest gardens are still common in the tropics and known by various names such as: *home gardens* in Kerala in South India, Nepal, Zambia, Zimbabwe and Tanzania; *Kandyan forest gardens* in Sri Lanka;[5] *huertos familiares*, the "family orchards" of Mexico; and *pekarangan*, the gardens of "complete design", in Java.[6] These are also called agroforests and, where the wood components are short-statured, the term shrub garden is employed. Forest gardens have been shown to be a significant source of income and food security for local populations.[7]

Robert Hart adapted forest gardening for the United Kingdom's temperate climate during the 1980s.[1] His theories were later developed by Martin Crawford from the Agroforestry Research Trust and various permaculturalists such as Graham Bell, Patrick Whitefield, Dave Jacke and Geoff Lawton.

9.2   In tropical climates

Forest gardens, or home gardens, are common in the tropics, using intercropping to cultivate trees, crops, and livestock on the same land. In Kerala in south India as well as in northeastern India, the home garden is the most common form of land use and is also found in Indonesia. One example combines coconut, black pepper, cocoa and pineapple. These gardens exemplify polyculture, and conserve much crop genetic diversity and heirloom plants that are not found in monocultures. Forest gardens have been loosely compared to the religious concept of the Garden of Eden.[8]

Robert Hart's forest garden in Shropshire

9.2.1 Americas

The BBC's *Unnatural Histories* claimed that the Amazon rainforest, rather than being a pristine wilderness, has been shaped by humans for at least 11,000 years through practices such as forest gardening and *terra preta*.[9] This was also explored in the bestselling book *1491* by author Charles C. Mann. Since the 1970s, numerous geoglyphs have also been discovered on deforested land in the Amazon rainforest, furthering the evidence about Pre-Columbian civilizations.[10][11]

On the Yucatán Peninsula, much of the Maya food supply was grown in "orchard-gardens", known as *pet kot*.[12] The system takes its name from the low wall of stones (*pet* meaning circular and *kot* wall of loose stones) that characteristically surrounds the gardens.[13]

9.2.2 Africa

In many African countries, for example Zambia, Zimbabwe, Ethiopia and Tanzania, gardens are widespread in rural, periurban and urban areas and they play an essential role in establishing food security. Most well known are the Chaga or Chagga gardens on the slopes of Mt. Kilimanjaro in Tanzania. These are an excellent example of an agroforestry system. In many countries, women are the main actors in home gardening and food is mainly produced for subsistence. In North-Africa, oasis layered gardening with palm trees, fruit trees and vegetables is a traditional type of forest garden.

9.2.3 Nepal

In Nepal, the *Ghar Bagaincha*, literally "home garden", refers to the traditional land use system around a homestead, where several species of plants are grown and maintained by household members and their products are primarily intended for the family consumption (Shrestha et al., 2002). The term "home garden" is often considered synonymous to the kitchen garden. However, they differ in terms of function, size, diversity, composition and features (Sthapit et al., 2006). In Nepal, 72% of households have home gardens of an area 2–11% of the total land holdings (Gautam et al., 2004). Because of their small size, the government has never identified home gardens as an important unit of food production and they thereby remain neglected from research and development. However, at the household level the system is very important as it is an important source of quality food and nutrition for the rural poor and, therefore, are important contributors to the household food security and livelihoods of farming communities in Nepal. The gardens are typically cultivated with a mixture of annual and perennial plants that can be harvested on a daily or seasonal basis. Biodiversity that has an immediate value is maintained in home gardens as women and children have easy access to preferred food. Home gardens, with their intensive and multiple uses, provide a safety net for households when food is scarce. These gardens are not only important sources of food, fodder, fuel, medicines, spices, herbs, flowers, construction materials and income in many countries, they are also important for the in situ conservation of a wide range of unique genetic resources for food and agriculture (Subedi et al., 2004). Many uncultivated, as well as neglected and underutilised species could make an important contribution to the dietary diversity of local communities (Gautam et al., 2004).

In addition to supplementing diet in times of difficulty, home gardens promote whole-family and whole-community involvement in the process of providing food. Children, the elderly, and those caring for them can participate in this infield agriculture, incorporating it with other household tasks and scheduling. This tradition has existed in many cultures around the world for thousands of years.[14][15]

9.3 In Mediterranean climates

The Mediterranean climate has long, hot, rainless summers and relatively short, cool, rainy winters (Köppen climate classification *Csa*).[16] Its climate conditions are highly variable within an area and modified locally by altitude, latitude, and the proximity to the Mediterranean.[16] In the 1950s the Forest Research Department of the Ministry of Agriculture founded a botanical forest garden in the Sharon region in Israel, the Ilanot Forest.[17] As the only one of its kind in Israel, it harbours more than 750 species of trees from locations all over the world, including the Japanese sago palm cycas revoluta, fig trees (ficus glomerata), stone pine trees (pinus pinea) that produce tasty pine nuts and adds to the biodiversity of Israel.

9.4   In temperate climates

Robert Hart, forest gardening pioneer

Robert Hart coined the term "forest gardening" during the 1980s. Hart began farming at Wenlock Edge in Shropshire with the intention of providing a healthy and therapeutic environment for himself and his brother Lacon.[18] Starting as relatively conventional smallholders, Hart soon discovered that maintaining large annual vegetable beds, rearing livestock and taking care of an orchard were tasks beyond their strength. However, a small bed of perennial vegetables and herbs he planted was looking after itself with little intervention.

Following Hart's adoption of a raw vegan diet for health and personal reasons, he replaced his farm animals with plants. The three main products from a forest garden are fruit, nuts and green leafy vegetables.[19] He created a model forest garden from a 0.12 acre (500 m^2) orchard on his farm and intended naming his gardening method *ecological horticulture* or *ecocultivation*.[20] Hart later dropped these terms once he became aware that *agroforestry* and *forest gardens* were already being used to describe similar systems in other parts of the world.[21] He was inspired by the forest farming methods of Toyohiko Kagawa and James Sholto Douglas, and the productivity of the Keralan home gardens as Hart explains:[22]

> From the agroforestry point of view, perhaps the world's most advanced country is the Indian state of Kerala, which boasts no fewer than three and a half million forest gardens...As an example of the extraordinary intensity of cultivation of some forest gardens, one plot of only 0.12 hectares (0.30 acres) was found by a study group to have twenty-three young coconut palms, twelve cloves, fifty-six bananas, and forty-nine pineapples, with thirty pepper vines trained up its trees. In addition, the small holder grew fodder for his house-cow.[23]

9.4.1 Seven-layer system

1. CANOPY (LARGE FRUIT & NUT TREES)
2. LOW TREE LAYER (DWARF FRUIT TREES)
3. SHRUB LAYER (CURRANTS & BERRIES)
4. HERBACEOUS (COMFREYS, BEETS, HERBS)
5. RHIZOSPHERE (ROOT VEGETABLES)
6. SOIL SURFACE (GROUND COVER, EG, STRAWBERRY, ETC)
7. VERTICAL LAYER (CLIMBERS, VINES)

THE FOREST GARDEN: A SEVEN LEVEL BENEFICIAL GUILD

The seven layers of the forest garden

Robert Hart pioneered a system based on the observation that the natural forest can be divided into distinct levels. He used intercropping to develop an existing small orchard of apples and pears into an edible polyculture landscape consisting of the following layers:

1. 'Canopy layer' consisting of the original mature fruit trees.

2. 'Low-tree layer' of smaller nut and fruit trees on dwarfing root stocks.

3. 'Shrub layer' of fruit bushes such as currants and berries.

4. 'Herbaceous layer' of perennial vegetables and herbs.

5. 'Rhizosphere' or 'underground' dimension of plants grown for their roots and tubers.

6. 'Ground cover layer' of edible plants that spread horizontally.

7. 'Vertical layer' of vines and climbers.

A key component of the seven-layer system was the plants he selected. Most of the traditional vegetable crops grown today, such as carrots, are sun loving plants not well selected for the more shady forest garden system. Hart favoured shade tolerant perennial vegetables.

9.4.2 Further development

The Agroforestry Research Trust (ART), managed by Martin Crawford, runs experimental forest gardening projects on a number of plots in Devon, United Kingdom.[24] Crawford describes a forest garden as a low-maintenance way of sustainably producing food and other household products.[25]

Ken Fern had the idea that for a successful temperate forest garden a wider range of edible shade tolerant plants would need to be used. To this end, Fern created the organisation Plants for a Future (PFAF) which compiled a plant database suitable for such a system. Fern used the term *woodland gardening*, rather than forest gardening, in his book *Plants for a Future*.[26][27]

The Movement for Compassionate Living (MCL) promote forest gardening and other types of vegan organic gardening to meet society's needs for food and natural resources. Kathleen Jannaway, the founder of MCL, wrote a book outlining a sustainable vegan future called *Abundant Living in the Coming Age of the Tree* in 1991. In 2009, the MCL provided a grant of £1,000 to the Bangor Forest Garden project in Gwynedd, North West Wales.[28]

Kevin Bradley coined the phrase "Edible Forest" in the 1980s as the name of his nursery, garden, and orchard on 5 acres in the frigid zone 3 pine forests of northern Wisconsin. Among 3 options, he chose "Edible Forest" because it "evokes at once an ethereal, spiritual, and magical image", of Disney- like "Forest of No Return"; of the biblical "Garden of Eden". This image was perfectly in line with his ongoing experiment begun in 1985 in what he calls a closed loop human environment, combining multi- story tree and field crop "garden/orchards" for maximum beauty and use of space, someday to be very useful in an ever shrinking world. "The name, at the same time, with its irrational first impression (of course we can't eat a forest), forces the mind to think, if just a little bit, about its inference and thus sticks in our memories". It appeared from Bradley's research that the two words had, prior to the 80's, never been put together before as a noun phrase but which by today, after more than two decades of Bradley's "Edible Forest Nursery" and the 2005 text by Jacke and Toensmeirer's- "Edible Forest Gardens", has grown into a movement and little "Edible Forests" all over the world.

In 2005, Dave Jacke and Eric Toensmeier's two-volume *Edible Forest Gardens* provided a deeply researched reference focused on North American forest gardening climates, habitats, and species. The book attempts to ground forest gardening deeply in ecological science. The Apios Institute wiki grew out of their work, and seeks to document and share the experience of people around the world working with the species in polycultures.

9.4.3 Permaculture

Bill Mollison, who coined the term *permaculture*, visited Robert Hart at his forest garden in Wenlock Edge in October 1990.[29] Hart's seven-layer system has since been adopted as a common permaculture design element.

Numerous permaculturalists are proponents of forest gardens, or food forests, such as Graham Bell, Patrick Whitefield, Dave Jacke, Eric Toensmeier and Geoff Lawton. Bell started building his forest garden in 1991 and wrote the book *The Permaculture Garden* in 1995, Whitefield wrote the book *How to Make a Forest Garden* in 2002, Jacke and Toensmeier co-authored the two volume book set *Edible Forest Gardening* in 2005, and Lawton presented the film *Establishing a Food Forest* in 2008.[30][31][32]

Austrian Sepp Holzer practices "Holzer Permaculture" on his *Krameterhof* farm, at varying altitudes ranging from 1,100 to 1,500 metres above sea level. His designs create micro-climates with rocks, ponds and living wind barriers, enabling the cultivation of a variety of fruit trees, vegetables and flowers in a region that averages 4 °C, and with temperatures as low as −20 °C in the winter.

9.5 Projects

El Pilar on the Belize-Guatemala border features a forest garden to demonstrate traditional Maya agricultural practices.[33][34] A further 1-acre model forest garden, called Känan K'aax (meaning well-tended garden in Mayan), is being funded by the National Geographic Society and developed at Santa Familia Primary School in Cayo.[35]

In the United States the largest known food forest on public land is believed to be the 7-acre Beacon Food Forest in Seattle, Washington.[36] Other forest garden projects include those at the Central Rocky Mountain Permaculture Institute in Basalt, Colorado and Montview Neighborhood farm in Northampton, Massachusetts.[37][38]

In Canada food forester Richard Walker has been developing and maintaining food forests in the province of British Columbia for over 30 years. He developed a 3-acre food forest that when at maturity provided raw materials for a nursery and herbalism business as well as food for his family.[39] The Living Centre have developed various forest garden projects in Ontario.[40]

In the United Kingdom, other than those run by the Agroforestry Research Trust (ART), there are numerous forest garden projects such as the Bangor Forest Garden in Gwynedd, North West Wales.[41] Martin Crawford from ART administers the Forest Garden Network, an informal network of people and organisations around the world who are cultivating their own forest gardens.[42][43]

9.6   See also

- Agroecology

- Analog forestry

- Climate-friendly gardening

- Deep ecology

- Forest farming

- Gardening

- Hügelkultur

- List of companion plants

- Mycoforestry

- Multiple cropping

- Natural farming

- Nutrient cycle

- Orchard

- Permaculture

- Polyculture

- Vegan organic gardening

9.7   Notes

[1] Crawford, Martin (2010). *Creating a Forest Garden*. Green Books. p. 18.

[2] Hart, Robert A. de J. (1996a), p.124: "Forest gardening, in the sense of finding uses for and attempting to cololntrol the growth of wild plants, is undoubtedly the oldest form of land use in the world."

[3] Douglas John McConnell (2003). *The Forest Farms of Kandy: And Other Gardens of Complete Design*, p.1, "Forest garden farms are probably the world's oldest and most resilient agroecosystem."

[4] Douglas John McConnell (1992). *The Forest-Garden Farms of Kandy, Sri Lanka*. p. 1. ISBN 9789251028988.

[5] Jacob, V. J.; Alles, W. S. (1987). "Kandyan gardens of Sri Lanka". *Agroforestry Systems*. **5** (2): 123. doi:10.1007/BF00047517.

[6] timeshighereducation.co.uk

[7] Douglas John McConnell (1973). *The economic structure of Kandyan forest-garden farms*.

[8] Graham Bell (2004). *The Permaculture Garden*, p.129, "The Forest Garden...This is the original garden of Eden. It could be your garden too."

- Also see Rob Hopkins (foreword), Martin Crawford (2010). *Creating a Forest Garden: Working with Nature to Grow Edible Crops*, p.10 "Perhaps what Hart created was the closest to what we imagine the Garden of Eden as being."

- Helmut Lieth (1989). *Tropical Rain Forest Ecosystems: Biogeographical and Ecological Studies*, p.611 "Important food plants, such as sago-producing palms, fruit-producing trees and medicinal plants were purposefully aggregated and tended in convenient places. Eventually, the forest garden, a kind of Garden of Eden, emerged. These jungle gardens on good soils of easy access required little maintenance and hardly any hard work."

- Dave Jacke and Eric Toensmeier (2005). *Edible Forest Gardens - Volume One*, p.1

- Robert Hart (1996a), p.80

- Deborha d'Arms (2011). *Jardin D'Or: A Treatise on Forest Gardening, Recreating Sustainable Gardens of Eden*

[9] "Unnatural Histories - Amazon". *BBC Four*.

[10] Simon Romero (January 14, 2012). "Once Hidden by Forest, Carvings in Land Attest to Amazon's Lost World". *The New York Times*.

[11] Martti Pärssinen, Denise Schaan and Alceu Ranzi (2009). "Pre-Columbian geometric earthworks in the upper Purús: a complex society in western Amazonia". *Antiquity*. **83** (322): 1084–1095.

[12] Michael Ernest Smith and Marilyn A. Masson (2000). *The Ancient Civilizations of Mesoamerica*. p. 127. ISBN 9780631211167.

[13] David L. Lentz, ed. (2000). *Imperfect Balance: Landscape Transformations in the Precolumbian Americas*. p. 212. ISBN 9780231111577.

[14] Killion, Thomas W., *Gardens of Prehistory: The Archaeology of Settlement Agriculture in Greater Mesoamerica*, University of Alabama Press, 1992

[15] Heidelberg, Kurt, "Ethnographic Analogy and Its Problems in the Northern Maya Lowlands". In *Lifeways in the Northern Maya Lowlands: New Approaches to Archaeology in the Yucatan Peninsula*. Edited by Jennifer Mathews. University of Arizona Press. 2006

[16] "Climate". U.S. Library of Congress. Retrieved 8 April 2008.

[17] "Ilanot Forest – A Botanical Forest Garden". *kkl.org.il*. KKL JNF. Retrieved 22 September 2015.

[18] Graham Burnett. "Seven Storeys of Abundance; A visit to Robert Hart's Forest Garden".

[19] Patrick Whitefield (2002). *How to Make a Forest Garden*. p. 5. ISBN 9781856230087.

[20] Hart, Robert A. de J. (1996a), p. 45

[21] Hart, Robert A. de J. (1996a), pages 28 and 43

[22] Hart, Robert A. de J. (1996a), p. 41

[23] Hart, Robert A. de J. (1996a), pages 4–5

[24] "Agroforestry Research Trust".

[25] "Forest gardening". Agroforestry Research Trust. Retrieved 13 Feb 2013.

[26] "Woodland Gardening".

[27] "Plants for a Future - The book".

[28] "Bangor Forest Garden" (PDF). The Movement for Compassionate Living - New Leaves (issue no.93). 2009: 6–8.

[29] Hart, Robert A. de J. (1996a), p. 149

[30] "Graham Bell's Forest Garden".

[31] "Edible Forest Gardening".

[32] "*Establishing a Food Forest* review".

[33] Ford, Anabel (May 2, 2009). "El Pilar Archaeological Reserve for Maya Flora and Fauna". *The Guatemala Times*. Retrieved 2009-07-26.

[34] Ford, Anabel (December 15, 2010). "Legacy of the Ancient Maya: The Maya Forest Garden". Popular Archaeology.

[35] "National Geographic Society Funds Mayan Garden".

[36] Mellinger, Robert (16 February 2012). "Nation's Largest Food Forest takes root on Beacon Hill". *Crosscut*. Retrieved 14 March 2012.

[37] "The Central Rocky Mountain Permaculture Institute".

[38] "Montview Neighborhood farm".

[39] "Richard Walker".

[40] "Forest Gardening".

[41] "Bangor Forest Garden".

[42] "The Agroforestry and Forest Garden Network".

[43] Martin Crawford (2014). "List of visitable forest garden and agroforestry projects in the UK, Europe and North America". Agroforestry Research Trust.

9.8   References

- Crawford, Martin 2010. *Creating a Forest Garden: Working with Nature to Grow Edible Crops*. Totnes: Green Books. ISBN 1-900322-62-5.

- d'Arms, Deborha 2011. *Jardin d'Or (Garden of Gold): A Treatise on Forest Gardening, Recreating Sustainable Gardens of Eden*. Los Gatos, CA: Robertson Publishing. ISBN 978-1611700299.

- Douglas, J. Sholto and Hart, Robert A. de J. 1985. *Forest Farming*. Intermediate Technology. ISBN 0-946688-30-3.

- Fern, Ken 1997. *Plants for a Future: Edible and Useful Plants for a Healthier World*. Hampshire: Permanent Publications. ISBN 1-85623-011-2.

- Hart, Robert A. de J. (1996a). *Forest Gardening: Cultivating an Edible Landscape*. White River Junction, VT: Chelsea Green. ISBN 0-930031-84-9.

- Hart, Robert A. de J. 1996b. *Beyond the Forest Garden*. Gaia Books. ISBN 1-85675-037-X.

- Jacke, Dave, and Toensmeier, Eric 2005. *Edible Forest Gardens*. Two volume set. Volume One: *Ecological Vision and Theory for Temperate Climate Permaculture*, ISBN 1-931498-79-2. Volume Two: *Ecological Design and Practice for Temperate Climate Permaculture*, ISBN 1-931498-80-6. White River Junction, VT: Chelsea Green.

- Jannaway, Kathleen 1991. *Abundant Living in the Coming Age of the Tree*. Movement for Compassionate Living. ISBN 0-9517328-0-3.

- Smith, Joseph Russell 1988 (first published in 1929). *Tree Crops: A Permanent Agriculture*. Island Press. ISBN 0-933280-44-0

- Whitefield, P. 2002. *How to Make a Forest Garden*. Hampshire: Permanent Publications. ISBN 1-85623-008-2.

9.9   External links

- Why Food Forests?, Permaculture Research Institute

- Plant an Edible Forest Garden, *Mother Earth News*

- The garden of the future?, *The Guardian*

- Edible Forest Gardens: an Invitation to Adventure, *The Natural Farmer*

- Forest gardens, Permaculture Association

- El Pilar Forest Garden Network, information on traditional Maya forest gardening

Chapter 10

Inga alley cropping

Inga alley cropping refers to planting agricultural crops between rows of Inga trees. It has been promoted by Mike Hands.[1]

Using the Inga tree for alley cropping has been proposed as an alternative to the ecological destruction of slash and burn cultivation.[2] The technique has been found to increase yields. It is sustainable agriculture as it allows the same plot to be cultivated over and over again thus eliminating the need for burning of the rainforests to get fertile plots.

10.1　Inga tree

The flower of an Inga tree.

Main article: Inga

Inga trees are native to many parts of Central and South America. Inga grows well on the acid soils of the tropical rainforest and former rainforest. They are leguminous and fix nitrogen into a form usable by plants.[3] Mycorrhiza growing within the roots (arbuscular mycorrhiza) was found to take up spare phosphorus, allowing it to be recycled into the soil.[4]

Other benefits of Inga include the fact that it is fast growing with thick leaves which, when left on the ground after pruning, form a thick cover that protects both soil and roots from the sun and heavy rain. It branches out to form a thick canopy so as to cut off light from the weeds below and withstands careful pruning year after year.[3]

10.2   History

The technique was first developed and trialled by tropical ecologist Mike Hands in Costa Rica in the late 1980s and early '90s. Research funding from the EEC allowed him to experiment with species of Inga. Although alley cropping had been widely researched, it was thought that the tough pinnate leaves of the Inga tree would not decompose quickly enough.[2]

As the crops grow, so does the Inga. When the crops are harvested the Inga is allowed to grow back. Once more it closes the canopy, is pruned, and the cycle is repeated, time and again. Leaves pruned from the tree decompose on the ground releasing phosphorus for crops. Fungi take up phosphorus to repeat the cycle.

The effects of slash-and-burn agriculture on primary forest, such as seen in this picture, can be devastating. Inga alley cropping may be a solution.

Using this system, not only do the farmers grow their basic crops of maize and beans, but also cash crops. Previously this was not possible because when the plot was a distance from the farmer's home, consistent guarding and tending could be too challenging. Now with the same plot being used continuously, it can be near home, thus allowing an entire family to help to tend and guard it, even when there are young children.

The Inga is used as hedges and pruned when large enough to provide a mulch in which bean and corn seeds are planted. This results in both improving crop yields and the retention of soil fertility on the plot that is being farmed. Hands had seen the devastating consequences that are caused by slash and burn agriculture while working in Honduras; this new technique seemed to offer the solution to the environmental and economic problems faced by so many slash and burn farmers.

Although this technique has the potential to save rainforest and lift many out of poverty, Inga alley cropping has not yet reached its full potential, although the charity Inga Foundation, headed by Mike Hands, has been consulted about potential projects in Haiti (which is almost completely deforested) and the Congo. Discussions have also been mooted about projects in Peru and Madagascar.

10.3 Method

For Inga alley cropping the trees are planted in rows (hedges) close together, with a gap, the alley, of about 4m between the rows. An initial application of rock phosphate has kept the system going for many years.

When the trees have grown, usually in about two years, the canopies close over the alley and cut off the light and so smother the weeds.

The trees are then carefully pruned. The larger branches are used for firewood. The smaller branches and leaves are left on the ground in the alleys. These rot down into a good mulch (compost). If any weeds haven't been killed off by lack of light the mulch smothers them.

The farmer then pokes holes into the mulch and plants his crops into the holes.

The crops grow, fed by the mulch. The crops feed on the lower layers while the latest prunings form a protective layer over the soil and roots, shielding them from both the hot sun and heavy rain. This makes it possible for the roots of both the crops and the trees to stay to a considerable extent in the top layer of soil and the mulch, thus benefiting from the food in the mulch, and escaping soil pests and toxic minerals lower down. Pruning the Inga also makes its roots die back, thus reducing competition with the crops.

10.4 Mike Hands

Mike Hands is a former academic researcher, current UK farmer and environmentalist, who pioneered the research into Inga alley cropping and currently promotes the concept through the Inga Foundation, a UK registered charity he founded in 2007 to support projects in Central and South America.

When the UK Environment Agency listed the "all-time" "scientists, campaigners, writers, economists and naturalists" who, in its view, have done the most to save the planet, Hands was placed 44th, just below Andrew Lees, and one place above German politician and activist, Petra Kelly.[5]

10.5 See also

- Shifting cultivation

10.6 References

[1] David Adam, *Earthshakers: the top 100 green campaigners of all time* The Guardian, 28 November 2006

[2] Elkan, Daniel. *Slash-and-burn farming has become a major threat to the world's rainforest* The Guardian 21 April 2004

[3] rainforestsaver.org: What is Inga alley cropping?

[4] Guinness, Bunny. *A "miracle cure"* The Sunday Telegraph 5 December 2004

[5] David Adam,*Earthshakers: the top 100 green campaigners of all time* The Guardian, 28 November 2006

- T.D.Pennignton and E.C.M. Fernandes (editors) "The Genus Inga, Utilization" Inga species and alley-cropping by Mike Hands, Kew Publications.

10.7   External links

- Inga Foundation

- Rainforest Saver Foundation (Inga alley cropping projects in Honduras and Cameroon)

- Inga alley cropping as an agrometeorogical service to slash and burn cultivation

- What is inga alley cropping?

Chapter 11

Intercropping

Intercropping is a multiple cropping practice involving growing two or more crops in proximity. The most common goal of intercropping is to produce a greater yield on a given piece of land by making use of resources that would otherwise not be utilized by a single crop.[1] Careful planning is required, taking into account the soil, climate, crops, and varieties. It is particularly important not to have crops competing with each other for physical space, nutrients, water, or sunlight. Examples of intercropping strategies are planting a deep-rooted crop with a shallow-rooted crop, or planting a tall crop with a shorter crop that requires partial shade. Inga alley cropping has been proposed as an alternative to the ecological destruction of slash-and-burn farming.[2]

When crops are carefully selected, other agronomic benefits are also achieved. Lodging-prone plants, those that are prone to tip over in wind or heavy rain, may be given structural support by their companion crop.[3] Creepers can also benefit from structural support. Some plants are used to suppress weeds or provide nutrients.[4] Delicate or light-sensitive plants may be given shade or protection, or otherwise wasted space can be utilized. An example is the tropical multi-tier system where coconut occupies the upper tier, banana the middle tier, and pineapple, ginger, or leguminous fodder, medicinal or aromatic plants occupy the lowest tier.

Intercropping of compatible plants also encourages biodiversity, by providing a habitat for a variety of insects and soil organisms that would not be present in a single-crop environment. This in turn can help limit outbreaks of crop pests by increasing predator biodiversity.[5] Additionally, reducing the homogeneity of the crop increases the barriers against biological dispersal of pest organisms through the crop.

There are several ways pests can be controlled through intercropping:

- **Trap cropping,** this involves planting a crop nearby that is more attractive for pests compared to the production crop, the pests will target this crop and not the production crop.

- **Repellant intercrops,** an intercrop that has a repellent effect to certain pests can be used. This system involved the repellant crop masking the smell of the production crop in order to keep pests away from it.

- **Push-pull cropping,** this is a mixture of trap cropping and repellant intercropping. An attractant crop attracts the pest and a repellant crop is also used to repel the pest away. [6]

The degree of spatial and temporal overlap in the two crops can vary somewhat, but both requirements must be met for a cropping system to be an intercrop. Numerous types of intercropping, all of which vary the temporal and spatial mixture to some degree, have been identified.[7][8] These are some of the more significant types:

- **Mixed intercropping,** as the name implies, is the most basic form in which the component crops are totally mixed in the available space.

- **Row cropping** involves the component crops arranged in alternate rows. Variations include alley cropping, where crops are grown in between rows of trees, and strip cropping, where multiple rows, or a strip, of one crop are alternated with multiple rows of another crop. A new version of this is to intercrop rows of solar photovoltaic modules with agriculture crops. This practice is called agrivoltaics.[9]

- **Temporal intercropping** uses the practice of sowing a fast-growing crop with a slow-growing crop, so that the fast-growing crop is harvested before the slow-growing crop starts to mature.

- Further temporal separation is found in **relay cropping**, where the second crop is sown during the growth, often near the onset of reproductive development or fruiting, of the first crop, so that the first crop is harvested to make room for the full development of the second.

11.1   See also

- Agrivoltaics

- Allotment garden

- Asset-based community development (ABCD)

- Community Food Security Coalition

- Community gardening

- Container garden

- Companion planting

- Ecological sanitation

- Food-feed system

- Forest gardening

- Gardening

- Green wall

- Monoculture

- Organic farming

- Permaculture

- Sustainable agriculture

11.2   References

[1] Ouma, George; Jeruto, P (2010). "Sustainable horticultural crop production through intercropping: The case of fruits and vegetable crops: A review" (PDF). *Agriculture and Biology Journal of North America.* **1** (5): 1098–1105.

[2] Elkan, Daniel. *Slash-and-burn farming has become a major threat to the world's rainforest* The Guardian 21 April 2004

[3] Trenbath, B.R. 1976. Plant interactions in mixed cropping communities. pp. 129–169 in R.I. Papendick, A. Sanchez, G.B. Triplett (Eds.), *Multiple Cropping.* ASA Special Publication 27. American Society of Agronomy, Madison, WI.

[4] Mt. Pleasant, Jane (2006). "The science behind the Three Sisters mound system: An agronomic assessment of an indigenous agricultural system in the northeast". In John E. Staller; Robert H. Tykot; Bruce F. Benz. *Histories of maize: Multidisciplinary approaches to the prehistory, linguistics, biogeography, domestication, and evolution of maize.* Amsterdam. pp. 529–537.

[5] Miguel Angel Altieri; Clara Ines Nicholls (2004). *Biodiversity and Pest Management in Agroecosystems, Second Edition.* Psychology Press.

[6] "Controlling Pests with Plants: The power of intercropping - UVM Food Feed". *UVM Food Feed*. 2014-01-09. Retrieved 2016-12-01.

[7] Andrews, D.J., A.H. Kassam. 1976. The importance of multiple cropping in increasing world food supplies. pp. 1–10 in R.I. Papendick, A. Sanchez, G.B. Triplett (Eds.), *Multiple Cropping*. ASA Special Publication 27. American Society of Agronomy, Madison, WI.

[8] Lithourgidis, A.S.; Dordas, C.A.; Damalas, C.A.; Vlachostergios, D.N. (2011). "Annual intercrops: an alternative pathway for sustainable agriculture" (PDF). *Australian Journal of Crop Science*. **5** (4): 396–410.

[9] Dinesh, Harshavardhan; Pearce, Joshua M. (2016-02-01). "The potential of agrivoltaic systems". *Renewable and Sustainable Energy Reviews*. **54**: 299–308. doi:10.1016/j.rser.2015.10.024.

[10] *Improving nutrition through home gardening*, Home Garden Technology Leaflet 13: Multilayer cropping, FAO, 2001

11.3 External links

- Intercropping at Washington State University

Chapter 12

Mycoforestry

Amanita *species are ectomycorrhizal with many trees*

Mycoforestry is an ecological forest management system implemented to enhance forest ecosystems and plant communities through the introduction of mycorrhizal and saprotrophic fungi. Mycoforestry is considered a type of permaculture[1] and can be implemented as a beneficial component of an agroforestry system. Mycoforestry can enhance the yields of tree crops and produce edible mushrooms, an economically valuable product. By integrating plant-fungal associations into a forestry management system, native forests can be preserved, wood waste can be recycled back into the ecosystem, planted restoration sites are enhanced, and the sustainability of forest ecosystems are improved.[2] Mycoforestry is an alternative to the practice of clearcutting, which removes dead wood from forests, thereby diminishing nutrient availability and reducing soil depth.[3]

12.1   Selection of fungal species

According to Paul Stamets, the first principle for the creation of a mycoforestry system is to utilize native fungal species. Implementing a mycoforestry system provides the potential of improving restoration efforts and the possibility of economic gain through mushroom cropping and harvesting. However to utilize native fungal flora, first the relationships between present fungal species and growth substrate, and habitat need to be studied.

A simple way to introduce a mycoforestry system and enhance out-plantings for crops and forest restoration sites is to "use mycorrhizal spore inoculum when replanting forest lands"[2] For this process it is best to match native trees with native mycorrhizal fungi. This method keeps and will promote the functioning of the native ecosystem, and native biodiversity.

It is assumed in a functioning forest ecosystem an underground mycelial network persists even if no fruiting bodies are visible.[4] A period of disappearance of mushrooms from an area should not cause alarm. In order to trigger the formation of fruiting bodies, many fungal species require specific environmental conditions. Most species of fungi do not fruit year round.

Mycoforestry is an emergent scientific field and practice.[2] Until broadly standardized protocols are created and perfected, the collection of both current and historical ecological site conditions will improve the success of the project.[2] Therefore, a survey of fungal relations at the site under both prime and poor conditions is beneficial to implementation of a mycoforestry system.

12.2   Saprotrophic fungi

The second principle is to promote saprotrophic fungi in the environment.[2] Saprophytic fungi are crucial to mycoforestry systems because these are the primary composers breaking down wood and returning nutrients to the soil for use by the rest of the forest ecosystem. This can be accomplished through inoculation of wood debris at site. Spored oils can be used in chainsaws when problematic or invasive hardwood requires felling. This method is a simple means to inoculate a tree. Additionally plug spawn can be implemented and injected into wood mass again prompting colonization by the selected fungus. Eventually repeated colonization efforts should not be necessary as many fungal life forms are strong and will spread and sustain in the soil on their own.[4]

In management of the mycoforestry system it is important that dead wood be in contact with the ground. This allows fungus to reach up from the soil and decompose fallen wood releasing nutrients at a much quicker rate then if the wood is left standing.[2] Additionally it is important to leave dead wood on site for decomposition back into the soil.[2] This philosophy is similarly based to the fact that clear cutting of a forest reduces soil nutrients and thickness.[3]

12.3   Beneficial fungal interactions

The third principal is to implement species known to benefit plant species.[2] These are commonly mycorrhizal fungus that form long term associations with plants, often extending inside of plants roots acting as an additional root system providing for better absorption of nutrients and water.

Utilizing mushroom species that attract insects could be a useful source of fish food. This practice makes the mycoforestry a larger system. Unlike most agriculture systems it helps the environment in a number of ways. It ties all biological aspects of the environment together, creating sustainable living and food production as well as sustainable fisheries similar to the ancient Hawaiian Ahupua'a, which utilized sustainable all portions of the land for environmental and food security.

Additionally fungal species can be implemented that compete with disease causing agents like Armillaria root rots[2] to provide long term protection of the forestry system.

Additionally, the implementation of an agroforestry system performs mycoremediation and mycofiltration activities cleaning up toxins and restoring the environment.

Edible oyster mushrooms (Pleurotus sp.) fruiting from a stump

12.4   See also

- Forestry portal

12.5   References

[1] Friedman, Zev. Digging In. *New Life Journal*. 1 May 2009.

[2] Stamets, Paul (2005). *Mycelium running: how mushrooms can help save the world*. Ten Speed Press. ISBN 1-58008-579-2.

[3] Dahlgren, R. A.; Driscoll, C. T. The effects of whole-tree clear-cutting on soil processes at the Hubbard Brook Experimental Forest, New Hampshire, USA. *Plant and Soil*. Volume 158, Number 2 / January 1994.

[4] Frankland, Juliet C. All you ever wanted to know about Mycelium. *NWFG Newsletter*. April 1997. (ISSN 1465-8054) Print.

12.6   External links

- Spinosa, Ron. Fungi and Sustainability. *Fungi magazine*. Spring 2008.
- Stamets, Paul. Mycotechnology. *Fungi Perfecti*.

Armillaria, *a parasitic fungus*

Chapter 13

Polyculture

Polyculture providing useful within-field diversity: companion planting of carrots and onions. The onion smell puts off carrot root fly, while the smell of carrots puts off onion fly.[1]

Polyculture is agriculture using multiple crops in the same space, providing crop diversity in imitation of the diversity of natural ecosystems, and avoiding large stands of single crops, or monoculture. It includes multi-cropping, intercropping, companion planting, beneficial weeds, and alley cropping. It is the raising at the same time and place of more than one species of plant or animal. Polyculture is one of the principles of permaculture.

13.1 Advantages

Polyculture, though it often requires more labor, has two main advantages over monoculture.

Polyculture reduces susceptibility to disease. For example, a study in China showed that planting several varieties of rice in the same field increased yields by 89%, largely because of a dramatic (94%) decrease in the incidence of disease, which made pesticides redundant.[2]

Polyculture increases local biodiversity. This is one example of reconciliation ecology, or accommodating biodiversity within human landscapes. This may also form part of a biological pest control program.

13.2 See also

- Agroecology

- Aquaponics

- Beneficial weeds

- Companion planting

- Forest gardening

- Heirloom plant

- Holistic management

- Home gardens

- Integrated Multi-trophic Aquaculture

- Monoculture

- Nurse crop

13.3 References

[1] "Companion Planting Guide". Thompson & Morgan. Retrieved 14 June 2016.

[2] Youyong Zhu; et al. (2000). "Genetic Diversity and Disease Control in Rice". *Nature* (406): 718–722. doi:10.1038/35021046.

13.4 External links

- Crop rotation and polyculture

- Polycultures in the Brazilian drylands

- Polyculture and disease prevention

- PolyCultures: Food Where We Live

- Integrated Polyculture Farming System

Chapter 14

Riparian buffer

A **riparian buffer** is a vegetated area (a "buffer strip") near a stream, usually forested, which helps shade and partially protect a stream from the impact of adjacent land uses. It plays a key role in increasing water quality in associated streams, rivers, and lakes, thus providing environmental benefits. With the decline of many aquatic ecosystems due to agriculture, riparian buffers have become a very common conservation practice aimed at increasing water quality and reducing pollution.

14.1 Benefits

Riparian buffers act to intercept sediment, nutrients, pesticides, and other materials in surface runoff and reduce nutrients and other pollutants in shallow subsurface water flow.[1] They also serve to provide habitat and wildlife corridors in primarily agricultural areas. They can also be key in reducing erosion by providing stream bank stabilization.

14.1.1 Water quality benefits

- Intercepting sediments/nutrients - Key to counteract eutrophication in downstream lakes and ponds which can be detrimental to aquatic habitats because of large fish kills that occur upon large-scale eutrophication.

- Intercepting pesticides - Riparian buffers keep chemicals that can be harmful to aquatic life out of the water. Some pesticides can be especially harmful if they bioaccumulate in the organism, with the chemicals reaching harmful levels once they are ready for human consumption.

- Bank stabilization - This is important because erosion can be a major problem in agricultural regions when cut (eroded) banks can take land out of production. Erosion can also lead to sedimentation and siltation of downstream lakes, ponds, and reservoirs. Siltation can greatly reduce the life span of reservoirs and the dams that create the reservoirs.

14.1.2 Habitat benefits

- Provide habitat - Riparian buffers can act as crucial habitat for a large number of species, especially those who have lost habitat due to agricultural land being put into production.

- Increase biodiversity - By adding this vegetated area of land near a water source it becomes a prime location for species that may have left the area due to non-conservation land use to re-establish. With this re-establishment the number of native species and biodiversity in general can be increased.

- Buffers acting as corridors - Buffers also serve a major role in wildlife habitat. The habitat provided by the buffers also double as corridors for species that have had their habitat fragmented by various land uses.

- Shading water - The large trees in the first zone of the riparian buffer provide shade and therefore cooling for the water, increasing productivity and increasing habitat quality for aquatic species.

- Large woody debris - When branches and stumps fall into the stream from the riparian zone, more stream habitat features are created. Carbon is added as an energy source for biota in the stream.

14.1.3 Economic benefits

- Increase land value - Often people who purchase land for recreational use are willing to pay more if there is more wooded area located on the land.

- Produce profitable alternative crops - Vegetation such as Black Walnut and Hazelnut, which can be profitably harvested, can be incorporated into the riparian buffer.

- Increase lease fees for hunting - The increased habitat means that the land will be more sought-after for hunting purposes.

14.2   Buffer design

A riparian buffer is usually split into three different zones, each having its own specific purpose for filtering runoff and interacting with the adjacent aquatic system. Buffer design is a key element in the effectiveness of the buffer. It is generally recommended that native species be chosen to plant in these three zones, with the general width of the buffer being 50 feet (15 m) on each side of the stream.[2]

Zone 1. This zone should function mainly to shade the water source and act as a bank stabilizer. The zone should include large native tree species that grow fast and can quickly act to perform these tasks. Although this is usually the smallest of the three zones and absorbs the fewest contaminants, most of the contaminants have been eliminated by Zone 2 and Zone 3.[3]

Zone 2. Usually made up of native shrubs, this zone provides a habitat for wildlife, including nesting areas for bird species. This zone also acts to slow and absorb contaminants that Zone 3 has missed. The zone is an important transition between grassland and forest.[3]

Zone 3. This zone is important as the first line of defense against contaminants. It consists mostly of native grasses and serves primarily to slow water runoff and begin to absorb contaminants before they reach the other zones. Although these grass strips should be one of the widest zones, they are also the easiest to install.[3]

Streambed Zone. The streambed zone of the riparian area is linked closely to Zone 1. Zone 1 provides fallen limbs, trees, and tree roots that in turn slow water flow, reducing erosional processes associated with increased water flow and flooding. This woody debris also increases habitat and cover for various aquatic species

The National Agroforestry Center has developed a Filter Strip Design tool (AgBufferBuilder), which is a GIS-based computer program for designing vegetative filter strips around agricultural fields that utilizes terrain analysis to account for spatially non-uniform runoff.

14.3   Species selection (example using both native Nebraska and introduced species)

In Zone 1: Cottonwood, Bur Oak, Hackberry, Swamp White Oak, Siberian Elm, Honeylocust, Silver Maple, Black Walnut, and Northern Red Oak.[4]

In Zone 2: manchurian apricot, Silver Buffaloberry, Caragana, Black Cherry, Chokecherry, Sandcherry, Peking Cotoneaster, Midwest Crabapple, Golden Currant, Elderberry, Washington Hawthorn, American Hazel, Amur Honeysuckle, Common Lilac, Amur Maple, American Plum, and Skunkbush Sumac.[4]

In Zone 3: Western Wheatgrass, Big Bluestem, Sand Bluestem, Sideoats Grama, Blue Grama, Hairy Grama, Buffalo Grass, Sand Lovegrass, Switchgrass, Little Bluestem, Indiangrass, Prairie Cordgrass, Prairie Dropseed, Tall Dropseed, Needleandthread, Green Needlegrass.

14.4 Managing forests in riparian area

Logging is sometimes recommended as a management practice in riparian buffers, usually to provide economic incentive. However, some studies have shown that logging can harm wildlife populations, especially birds. A study by the University of Minnesota found that there was a correlation between the harvesting of timber in riparian buffers and a decline in bird populations.[5] Therefore, logging is generally discouraged as an environmental practice, and left to be done in designated logging areas.

14.5 Conservation incentives

The Conservation Reserve Program (CRP), a farming assistance program in the United States, provides many incentives to landowners to encourage them to install riparian buffers around water systems that have a high chance of non-point water pollution and are highly erodible. For example, the Nebraska system of Riparian Buffer Payments offers the following:

- Annual rental payments = $62 to $116 per acre/year (up to $96 per acre per year for certain marginal pasture land)
- Sign up bonus: up-front payment of $10 per acre/year for the life of the contract (Example: $150/ac for a fifteen-year contract)
- Payments for 90% of costs for trees, shrubs, grass seed, site preparation and planting
- 10 to 15 year contracts
- Eligible areas:
  - Up to 180 feet (55 m) average width along each side of a continuous or intermittent stream
  - Strips of grass, shrubs, and/or trees can be used
  - Planting done by the local Natural Resource Districts (NRDs)

These incentives are offered to agriculturists to compensate them for their economic loss of taking this land out of production. If the land is highly erodible and produces little economic gain, it can sometimes be more economic to take advantage of these CRP programs.[6]

14.6 Effectiveness

Riparian buffers have undergone much scrutiny about their effectiveness, resulting in thorough testing and monitoring. A study done by the University of Georgia, conducted over a nine-year period, monitored the amounts of fertilizers that reached the watershed from the source of the application. It found that these buffers removed at least 60% of the nitrogen in the runoff, and at least 65% of the phosphorus from the fertilizer application. The same study showed that the effectiveness of the Zone 3 was much greater than that of both Zone 1 and 2 at removing contaminants.[7]

14.7 Long-term sustainability

After the initial installation of the riparian buffer, relatively little maintenance needs to be performed to keep the buffer in good condition. Once the trees and grasses reach maturity, they regenerate naturally and make a more effective buffer.

The sustainability of the riparian buffer makes it extremely attractive to landowners, since they do relatively little work and still receive payments. Riparian buffers have the potential to be the most effective ways to protect aquatic biodiversity, water quality and manage water resources in developing countries that lack the funds to install water treatment and supply systems in midsize and small towns.

14.8 See also

- Agricultural wastewater treatment

- Agroforestry

- Ecoscaping

- Erosion control

- Nonpoint source pollution

14.9 References

[1] U.S. Natural Resources Conservation Service (NRCS). (2006). "National Conservation Practice Standard: Riparian Forest Buffer." Code 391. January 2006.

[2] Dosskey, M., Schultz, D., & Isenhart, T. (1997). "Riparian Buffers for Agricultural Land." *Agroforestry Notes,* No. 3, January 1997. National Agroforestry Center, USDA Forest Service, Lincoln, NE.

[3] Maryland Cooperative Extension. "Riparian Forest Buffer Design, Establishment, and Maintenance." University of Maryland, 1998.

[4] Nebraska Association of Resources Districts (2003). "Conservation Trees for Nebraska."

[5] Journal of Wildlife Management; Apr 2005, Vol. 69 Issue 2, p689-698, 10p

[6] University of Nebraska Cooperative Extension. "Benefits of Riparian Forest Buffers (Streamside Plantings of Trees, Shrubs and Grasses)." University Press, Lincoln, NE.

[7] Durham, Sharon. "Riparian Buffers Effective." Southeast Farm Press. 4 Feb 2004. p26

14.10 External links

- National Agroforestry Center (USDA)

- Filter Strip Design Tool (USDA)

- Extensive Riparian Buffer bibliography

A riparian buffer of vegetation lining a farm creek in Story County, Iowa.

Ground level view of riparian buffer between Munson Pond (off camera left) and an agricultural operation (off camera right)

Chapter 15

Silvopasture

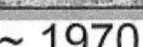

Silvopasture over the years.

Silvopasture (Latin, *silva* forest) is the practice of combining forestry and grazing of domesticated animals in a mutually beneficial way. Advantages of a properly managed silvopasture operation are enhanced soil protection and increased long-term income due to the simultaneous production of trees and grazing animals. The trees are managed for high-value sawlogs and, at the same time, provide shade and shelter for livestock and forage, reducing stress and sometimes increasing forage production.[1]

Perhaps the oldest agroforestry system used in the temperate regions of the world, silvopastoral systems are characterized by integrating trees with forage and livestock production. Such systems have the potential to increase agricultural production in the long term.

Silvopastoral systems are definitely the most prominent agroforestry practice in the United States, particularly in the southeast.

In the UK, there is a nationwide experiment for silvopastoral systems in which a number of tree species and planting densities are being studied over a range of sites. It is called The Silvopastoral National Network Experiment.[2]

Their experience shows that sheep use the trees for shelter from wind. This could provide significant animal welfare benefits. However the fact that the sheep do spend time close to trees results in greater soil compaction close to trees with the greatest compaction when trees are planted at very low densities. It is recommend that trees are planted at no less than 400 per hectare to ensure good establishment.

15.1   See also

- Browsing (herbivory)

- Dehesa

15.2   References

[1] "Silvopasture". *National Agroforestry Center*. National Agroforestry Center. 10 April 2015. Archived from the original on 19 August 2015. Retrieved 29 April 2015.

[2] http://www.agroforestry.ac.uk/systems/pastoral/snne03.html

- Venator, Charles R., Jurgen Glaeser and Reynaldo Soto. 1992. "A Silvopastoral Strategy" in Development or Destruction: The Conversion of Tropical Forest to Pasture in Latin America. pp. 281–292. Westview Press/Boulder

- The Farm Woodland Forum - Silvopastoral Agroforestry

15.3   External links

- National Agroforestry Center (USDA)

- Farm Woodland Forum - Silvopastoral Agroforestry

- The short film *Agroforestry Practices - Silvopasture (2004)* is available for free download at the Internet Archive

Chapter 16

Trees 4 Children

Trees 4 Children volunteers plant melia trees in Kenya.

Trees 4 Children is a cooperative charitable effort between the Make A Mark Foundation and the University of Wisconsin-Stevens Point, in which forestry principles are utilized to provide an investible model for economic growth through land development.

The organization is the brainchild of American entrepreneur and philanthropist John M. Noel. In 2006, Noel's Make A Mark Foundation partnered with the UWSP College of Natural Resources in an attempt to apply business solutions to sustainability and subsistence problems in Kenya.[1]

Since its inception, the non-profit organization has been focused on growing and processing trees for lumber to help contribute to the financial self-sustainability of the Nyumbani Village in Kitui, Kenya, a village which supports more than 1,000 HIV/AIDS-affected children and elders.[2] Noel worked in conjunction with local officials to persuade the Kenyan government to donate land to the village in which to plant. Trees 4 Children works with villagers and students and faculty from the UWSP College of Natural Resources to produce innovative crop planning and planting strategies. The students and teachers analyze each year's results and can then suggest changes to the crop selection, planting frequency, and planting and harvesting techniques for maximum yield.

In Nyumbani, Trees 4 Children has helped the villagers plant more than 50,000 trees on the village's 700 acres.[3] The goal is to plant 600 trees per acre, with 75 percent of the trees being melia,[4] a fast-growing relative of mahogany. Senna, leucaena, and several species of acacia are planted for timber, nitrogen fixation and fuel, while sapota, java plum, tamarind, and guava are planted for fruit.

In 2012 and 2013 UWSP students worked with villagers on the development of food forests. In the food forest, manure and daily irrigation are utilized to grow vegetables and encourage tree growth, while the partial shade and windbreak provided by the trees boosts yields of kale, carrots, onions, cilantro, peppers, pumpkins, passionfruit, and papayas.

Food and wood not used by the village can be sold, and the proceeds used for village improvements.

Rows of trees and crops planted by Trees 4 Children volunteers.

Each year, Trees 4 Children plants 30,000 new Melia trees at Nyumbani. Once mature, these trees will generate an annual profit of over $700,000 (based on current monetary value) through harvesting and selling the lumber, which will exceed the operating costs of village. The first crop of trees was planted in 2008, and will be ready for harvest in 2018. Until then, villagers will sell the seedlings of the trees to neighboring villages to raise money and awareness for the project.

With the initial harvest of Melia trees several years away, Trees 4 Children is also working in Nyumbani on the Sunflower Project. Sunflowers are being grown for cooking oil to eliminate villagers' dependence on purchased oil, and to generate revenue from sales of excess oil.

For the next phase of the project, Trees 4 Children will help villagers teach residents of surrounding areas how to replicate the agroforestry initiative. Long-term goals are to standardize processes and techniques and to expand the initiative across Africa and Asia.

16.1   References

[1] [Trees 4 Children: Using Sustainable Agroforestry to Battle African HIV/AIDS Pandemic. "Trees 4 Children: Using Sustainable Agroforestry to Battle African HIV/AIDS Pandemic"] Check |url= value (help). dogoodtripping.com. Retrieved September 13, 2010.

[2] https://www.youtube.com/watch?v=NQk-alykjQc

[3] "Trees 4 Children Project Empowers a Kenyan Village of AIDS Survivors to Become Self-Sufficient". Yahoo! Finance. Retrieved September 18, 2012.

[4] https://www.youtube.com/watch?v=X_J0y6p_9XM

Chapter 17

Windbreak

Aerial view of field windbreaks in North Dakota

A **windbreak** or **shelterbelt** is a plantation usually made up of one or more rows of trees or shrubs planted in such a manner as to provide shelter from the wind and to protect soil from erosion. They are commonly planted around the edges of fields on farms. If designed properly, windbreaks around a home can reduce the cost of heating and cooling and save energy. Windbreaks are also planted to help keep snow from drifting onto roadways and even yards.[1] Other benefits include providing habitat for wildlife and in some regions the trees are harvested for wood products.

Windbreaks and intercropping can be combined in a farming practice referred to as alleycropping. Fields are planted in rows of different crops surrounded by rows of trees. These trees provide fruit, wood, or protect the crops from the wind. Alley cropping has been particularly successful in India, Africa, and Brazil, where coffee growers have combined farming

One of the original buildings at Svappavaara, designed by Ralph Erskine, which forms a long windbreak.

and forestry.[2]

A further use for a shelterbelt is to screen a farm from a main road or motorway. This improves the farm landscape by reducing the visual incursion of the motorway, mitigating noise from the traffic and providing a safe barrier between farm animals and the road.

The term "windbreak" is also used to describe an article of clothing worn to prevent wind chill. Americans tend to use the term "windbreaker" whereas Europeans favor the term "windbreak".

Fences called "windbreaks" are also used. Normally made from cotton, nylon, canvas, and recycled sails, windbreaks tend to have three or more panels held in place with poles that slide into pockets sewn into the panel. The poles are then hammered into the ground and a windbreak is formed. Windbreaks or "wind fences" are used to reduce wind speeds over erodible areas such as open fields, industrial stockpiles, and dusty industrial operations. As erosion is proportional to wind speed cubed a reduction of wind speed of 1/2 (for example) will reduce erosion by over 80%.

17.1   Windbreak aerodynamics

In essence, when the wind encounters a porous obstacle such as a windbreak or shelterbelt, air pressure increases (loosely speaking, air *piles up*) on the windward side and (conversely) air pressure decreases on the leeward side. As a result, the airstream approaching the barrier is retarded, and a proportion of it is displaced up and over the barrier, resulting in a *jet* of higher wind speed aloft. The remainder of the impinging airstream, having been retarded in its approach, now circulates through the barrier to its downstream edge, pushed along by the decrease in pressure across the shelterbelt's width; emerging on the downwind side, that airstream is now further retarded by an adverse pressure gradient, because in the lee of the barrier, with increasing downwind distance air pressure *recovers* again to the ambient level. The result is that minimum wind speed occurs not at or within the windbreak, nor at its downwind edge, but further downwind - nominally, at a distance of about 3 to 5 times the windbreak height H. Beyond that point wind speed recovers, aided by downward momentum transport from the overlying, faster-moving stream. From the perspective of the Reynolds-averaged Navier–Stokes equations these effects can be understood as resulting from the loss of momentum caused by the drag of leaves and branches and would be represented by the body force f_i (a distributed momentum sink).[3]

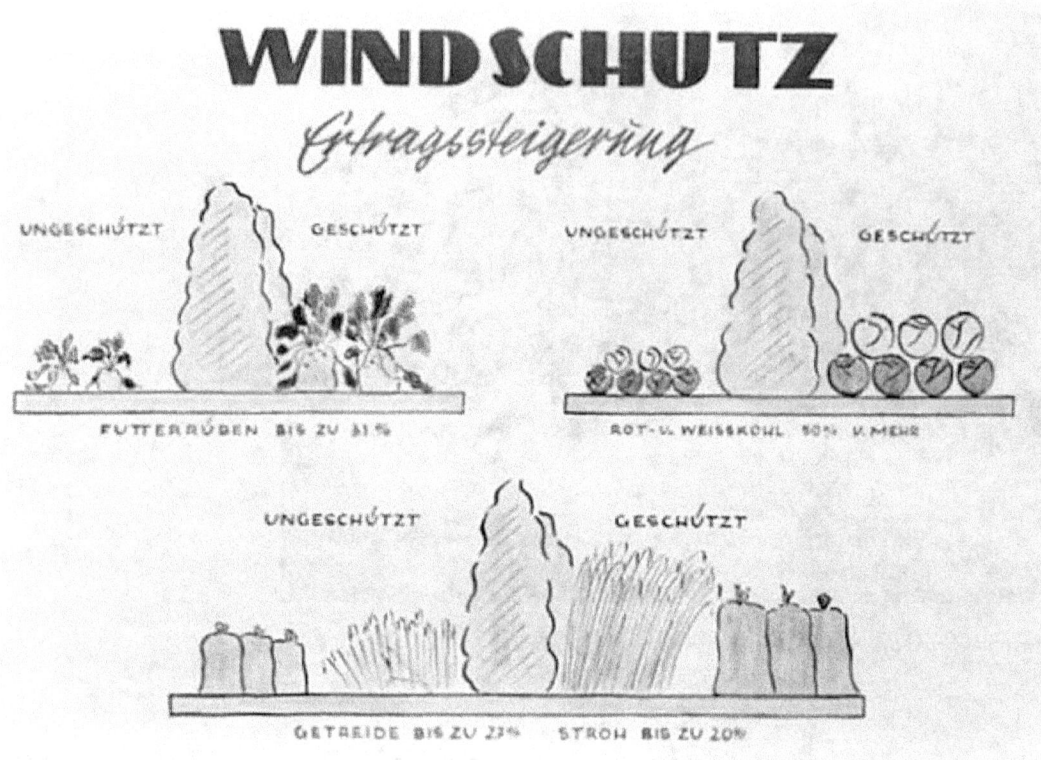

An East German windbreak promotion poster, 1952

Not only is the mean (average) wind speed reduced in the lee of the shelter, the wind is also less gusty, for turbulent wind fluctuations are also damped. As a result, turbulent vertical mixing is weaker in the lee of the barrier than it is upwind, and interesting secondary microclimatic effects result. For instance, by day sensible heat rising from the ground due to the absorption of sunlight (see *surface energy budget*) is mixed upward less efficiently in the lee of a windbreak, with the result that air temperature near ground is somewhat higher in the lee than on the windward side. Of course this effect is attenuated with increasing downwind distance and indeed, beyond about $8H$ downstream a zone may exist that is actually *cooler* than upwind.[4]

17.2   See also

- Agroforestry
- Buffer strip
- Dead hedge
- Desertification
- Energy-efficient landscaping
- Erosion control
- Great Plains Shelterbelt
- Hedgerow

- Macro-engineering

- Sahara Forest Project

- Sand fence

- Seawater Greenhouse

17.3   References

[1] "Windbreaks". *National Agroforestry Center*. Retrieved 29 April 2015.

[2] Withgott & Brennan 2008, p.249

[3] Wilson, 1985; Journal of Wind Engineering & Industrial Aerodynamics, Vol. 21

[4] Argete & Wilson, 1989, Agricultural & Forest Meteorology, Vol 48

Wilson, 1985; Journal of Wind Engineering & Industrial Aerodynamics, Vol. 21 Argete & Wilson, 1989, Agricultural & Forest Meteorology, Vol 48

17.4   Bibliography

- Withgott, Jay; Scott Brennan (2008). *Environment: The Science Behind the Stories* (3rd ed.). San Francisco, California: Pearson Benjamin Cummings. ISBN 0131357050.

17.5   External links

- National Agroforestry Center (USDA)

Chapter 18

World Agroforestry Centre

The **World Agroforestry Centre** (known as the International Council for Research in Agroforestry, ICRAF before 1991), is an international institute headquartered in Nairobi, Kenya, and founded in 1978. The Centre specializes in the sustainable management, protection and regulation of tropical rainforest and natural reserves. It is one of 15 agricultural research centres which makes up the global network known as the CGIAR (Consultative Group on International Agricultural Research). The Centre conducts research in agroforestry, in partnership with national agricultural research systems with a view to developing more sustainable and productive land use. The focus of its research is countries/regions in the developing world, particular in the tropics of Central and South America, Southeast Asia and parts of central Africa. In 2002 the Centre acquired the World Agroforestry Centre brand name, although **International Centre for Research in Agroforestry** remains its legal name and it continues to use the acronym **ICRAF**.

18.1   Missions

18.2   See also

- Forest Day

18.3   References

[1] World Agroforestry Blog. Retrieved 15/07/2015.

18.4   External links

- World Agroforestry Centre

Chapter 19

Paul Yeboah

Paul Yeboah, is an educator, farmer, permaculturist, community developer, and social entrepreneur. He is the founder and coordinator of the Ghana Permaculture Institute and Network in Techiman, Ghana, West Africa. It is located in the Brong-Ahafo Region of Ghana. The purpose of the Institute is to build and maintain a stable food system, to take care of the local ecosystems, and to improve the quality of life in the rural areas.[1][2] The GPN trains students and community in sustainable ecological farming techniques. They support projects through out Ghana; women groups, micro-finance projects; teach growing moringa; mushroom production; alley cropping, food forests development and Agroforestry.[3]

Permaculture is based on natural sustainable design systems. A agricultural system that uses practices to keep soil fertile, crops and livestock healthy. It encourages protection of the environment and an environmental lifestyle; so as to maintain environmental stability and maintain environmental resources for the future. It rehabilitates eroded and deforested land. The Permaculture Network encourages the practice of permaculture at home. The Permaculture Network's mission is to encourage, educate, and promote the use of permaculture by farmers and people in Ghana, which will contribute to the environmental soundness, and stability of the country's future.

They host international volunteers, interns, and students.[4] The Ghana Permaculture Network and Institute is a member of the Ghana Ecovillage Network. Which is an organization of sustainable development leaders and projects. Paul Yeboah is Vice President of the GEN which works towards promoting Indigenous Initiatives and Sustainability in Ghana. Permaculture is transforming communities in Ghana through education, food production, outreach, skills development, self-sufficiency, and creating small business enterprises.[5][6]

19.1 Background

At the age of 22 Paul Yeboah was concerned with rural and urban poverty. He received an Agricultural Certificate from the Farm Institute in Ghana. He initiated a rural and urban fruit forest project by using seed supplies from the Kade Oil Palm Research Institute and Bonsu Cocoa Research Station. The seeds and seedlings were given to the farmers on a credit basis. This project was instrumental in the creation of rural processing businesses and employment for the poor.

In 2003 Paul Yeboah was the farm manager for the Abbott of Kristo Buase Benedictine Monastery in Ghana. Greg Knibbs was invited to come to the Monastery to assist in the redesigning of the farm using Permaculture practices to restore the soil to fertility. The soil was depleted from the use of synthetic chemical pollution. Yeboah met Greg Knibbs and they worked together to form the Ghana Permaculture Network which later became the Ghana Permaculture Institute.

19.2 Career

The Ghana Permaculture Network was coordinated by Paul Yeboah in 2003. The GPN started out as a small farm demonstration training site that later grew into the Ghana Permaculture Institute (GPI). In 2007 the Ghana Permaculture Nwodua Tree Nursery was created. It was created to build community income, and to deal with environmental issues such

Ghana Brong Ahafo Region

as desertification and erosion. It also does reforestation. The work is collaborative community effort involving women, youth, and men. All members of the community are a part of the profits and benefits of working towards enhancing their environment. The project builds environmental awareness.[7]

Paul Yeboah is the Vice Chairman of the Ghana Ecovillage Network which was founded in 2012, and incorporated in 2013. It was formed by community leaders and groups with sustainable projects to promote Ecovillage strategies as models for sustainable development in Ghana.[8]

The Ghana Permaculture Network started out supporting local schools, community farmers in establishing tree nurseries, and tree planting projects. The Ghana Permaculture Network has now expanded in various parts of Ghana, into Togo, and Burkina Faso in West Africa.

A photograph of Tree Nursery in Techiman, Ghana.

19.3   Permaculture Institute Projects (Educational Programs)

The goal of the educational initiatives and projects is to promote lifelong learning with emphases on vocational training, and the need of viable skill development using practices of sustainable development.

- Permaculture Design Courses are taught in Ghana, Togo and Burkina Faso, West Africa.

- Ghana Permaculture Nwodua Tree Nursery Project - Ghana is a deforested country. The Tree Nursery Project addresses issues of deforestation in order to reverse erosion. It teaches and advocates the importance of trees and climate change issues.

- Oyster mushroom production - Using sawdust in bags the Permaculture Institute teaches how to grow mushrooms. This project allows participants to learn how to make a waste product sawdust useful. Which promotes healthy diet, nutrition and skill development.

- Demonstration Training sites for how-to build Permaculture home gardens

- Moringa Production - Which helps to facilitate the development of small-businesses to generate income. A project that promotes healthy nutrition and skills development.

- Ecovillage Design Course - Which teaches and acquaints students with tools that can be utilized to redevelop their communities in ecological economically, culturally and socially ways to foster sustainability.

Ghana Permaculture Network - Mushroom Production

19.4 External links

- Paul Yeboah Permaculture Institute

- Permaculture Design Course, Techiman Ghana

- Permaculture Design Course (Part two)

- Permaculture (What is Permaculture?)

- Ghana Permaculture Video Clips

- Permaculture and mushroom cultivation in Ghana

- Adele Women Association 'Upper Volta' Region Ghana

- Gaviotas is a model village of sustainable development in Colombia, South America.

- Urban Sustainable Development in Curitiba, Columbia

- Permaculture Institute - Alley Cropping and Food Forest 0001

Mushroom Vendor, Ghana

- Paul Yeboah on Panel Discussion about Permaculture.

- Permaculture and mushroom cultivation in Ghana

- Future In Our Hands International Network

19.5   Further Reading

- http://www.agriculturesnetwork.org/resources/extra/final-report-de-schutter

- Permaculture in Africa

- UN: Eco-Farming Feeds the World

19.6   References

[1] https://permacultureghana.wordpress.com/what-is-permaculture-2/

[2] http://edge5.com.au/permaculture-courses-fremantle/

[3] http://fiohnetwork.org/wp-content/uploads/2015/09/PAUL-TESTIMONY.pdf

[4] http://www.ghanaweb.com/GhanaHomePage/economy/artikel.php?ID=123567

[5] http://gen.ecovillage.org/en/gen-ghana

[6] https://www.permaculture.co.uk/how-permaculture-is-transforming-ghana

[7] http://permaculturenews.org/2011/11/01/the-ghana-permaculture-nwodua-tree-nursery-project-saving-lives-granting-livelihoods-and-restoring-

[8] http://gen.ecovillage.org/en/gen-ghana

19.7 Text and image sources, contributors, and licenses

19.7.1 Text

- **Agroforestry** *Source:* https://en.wikipedia.org/wiki/Agroforestry?oldid=751452968 *Contributors:* DavidLevinson, Kku, Robbot, Ddstretch, Alan Liefting, Bobblewik, Beland, Karol Langner, Rich Farmbrough, Vsmith, Bender235, CanisRufus, Smalljim, Eric Kvaalen, Paleorthid, Howrealisreal, Velella, Bobrayner, Mindmatrix, Tabletop, Firien, Bluemoose, Sphinxie, Mud4t, BD2412, Tlroche, Bruce1ee, Salix alba, Bedrupsbaneman, Boccobrock, DVdm, Sceptre, Pigman, Son of Paddy's Ego, Dialectric, Nirvana2013, VIGNERON, Brandon, Marcelo-Silva, LeonardoRob0t, SmackBot, EncycloPetey, Pushpam, Hmains, Bluebot, Hibernian, Brimba, Ggpauly, Salamurai, Byelf2007, Dmwilliams, Silk-Tork, Gobonobo, IronGargoyle, Rkmlai, Optimale, Caiaffa, Hu12, The Giant Puffin, Dan1679, Tahirs, Yaris678, A876, Gogo Dodo, Crowish, Shirulashem, Wawny, ThisIsAce, Liquid-aim-bot, Smartse, PhJ, Lfstevens, Ecoconservant, The Transhumanist, Zanzor, Gorav, Engineman, Pawl Kennedy, Sustainableyes, Gomm, Jeannie kendrick, Robin S, Keith D, Fincaproject, Skier Dude, Grmanners, AntiSpamBot, Entropy, Scott Roy Atwood, Burzmali, DASonnenfeld, TXiKiBoT, Scilit, Doug, HopsonRoad, Baf87, The Thing That Should Not Be, Mookie25, Pix- elBot, John Nevard, Abrech, Dana boomer, SoxBot III, DumZiBoT, XLinkBot, Rror, BiobulletM, K.t.1980, Lu Wunsch-Rolshoven, Addbot, Mgoldmo, Cr0, MrOllie, Granitethighs, Lightbot, Drpickem, Luckas-bot, Yobot, AnomieBOT, Afagitator, Jim1138, Minnecologies, Materi- alscientist, Citation bot, Wcoole, Sanja565658, RibotBOT, FrescoBot, Questionthedominantparadigm, W Nowicki, DrilBot, Pat604, Isiaunia, Tom.Reding, Manasij, Orenburg1, Trappist the monk, Theo10011, Tcazes, RjwilmsiBot, Look2See1, Zollerriia, Mmeijeri, ZéroBot, Don- ner60, Wormke-Grutman, 28bot, Petrb, ClueBot NG, Gareth Griffith-Jones, Mesoderm, Aghx, Kerrplunk, Helpful Pixie Bot, Gob Lofa, Island Monkey, Northamerica1000, Tom Pippens, Parvathisri, Suchthekaitlin, Rowan Adams, C.peterson32, Torontowiki, ChrisGualtieri, Lugia2453, Sidelight12, Laurelie237, Habibibibalani, AnuSingh855, Stamptrader, GoMinU, 7Sidz, Olenyash, Luke Smith232, MonzaMan09, QueenFan, Aidendonoghue, Jbanegas, Cailynjhkim, Anna Mayerhof, Lopezospina, 01234abcd, Great another anon, InternetArchiveBot, Ronaldkubo, Vladimirb777, GreenC bot and Anonymous: 103

- **Nutrecul Agroforestry Project** *Source:* https://en.wikipedia.org/wiki/Nutrecul_Agroforestry_Project?oldid=739755247 *Contributors:* Ma- gioladitis, LilHelpa, Orenburg1, BG19bot, Guy lemal, Narky Blert and Anonymous: 2

- **Agroforestry Research Trust** *Source:* https://en.wikipedia.org/wiki/Agroforestry_Research_Trust?oldid=538980017 *Contributors:* Mervyn, Nirvana2013, DASonnenfeld, Peter James and Lord Dandy

- **Beacon Food Forest** *Source:* https://en.wikipedia.org/wiki/Beacon_Food_Forest?oldid=748023361 *Contributors:* Timrollpickering, Alan Liefting, Brianhe, Dennis Bratland, Rjwilmsi, Tedder, Wavelength, Nirvana2013, TenPoundHammer, Dandelion1, Cydebot, Jdmarshall, Jllm06, The Anomebot2, Blacksqr, DASonnenfeld, Life of Riley, Arxiloxos, Yobot, Anna Frodesiak, Lumpytrout, Jonesey95, Skyerise, Kibi78704, AvicBot, Arcandam, Mogism, Shiningroad, Stamptrader, EChastain, Cahuitajacquelini, InternetArchiveBot and Anonymous: 2

- **Dehesa** *Source:* https://en.wikipedia.org/wiki/Dehesa?oldid=729642341 *Contributors:* William Avery, Jmabel, Alan Liefting, Eggstasy, Chan- heigeorge, SmackBot, Zeorymer, SmokeyJoe, Alanmaher, Nick Number, Escarbot, Steven Walling, DASonnenfeld, Richard New Forest, Hugo999, Pare Mo, Lightmouse, TubularWorld, No such user, SchreiberBike, MatthewVanitas, Addbot, HerculeBot, Yobot, AnomieBOT, Xufanc, Minnecologies, Apothecia, Madrid Tiger, Daouuud, Jenks24, Lynnstarrs, Imaginibus, Rowan Adams, Khazar2, Robert4565, Edwin- here, Dehesas ibericas, Asilah1981 and Anonymous: 13

- **Educational Concerns for Hunger Organization** *Source:* https://en.wikipedia.org/wiki/Educational_Concerns_for_Hunger_Organization? oldid=712165738 *Contributors:* Cydebot, Postcard Cathy, Ahusni, Onel5969, Fuhvah, BG19bot, Anne Delong and Anonymous: 1

- **Farm Forestry Toolbox** *Source:* https://en.wikipedia.org/wiki/Farm_Forestry_Toolbox?oldid=747570024 *Contributors:* GünniX, Jodi.a.schneider, DASonnenfeld, WereSpielChequers, Revent, Dodger67, Niceguyedc, MatthewVanitas, Yobot, BG19bot, Sulfurboy, Ekips39, PAT6592, Jhon- james83rock, JamesP and Anonymous: 3

- **Forest farming** *Source:* https://en.wikipedia.org/wiki/Forest_farming?oldid=710024036 *Contributors:* Kku, Skysmith, Lumos3, Alan Lief- ting, Bobo192, RJFJR, RHaworth, Nirvana2013, SmackBot, Derek R Bullamore, Bwpach, Bddmagic, Alaibot, Richhoncho, Fabrictramp, Gabriel Kielland, Gomm, Jeannie kendrick, DASonnenfeld, JL-Bot, Occur Curve, ClueBot, Auntof6, Aprock, MrOllie, AnomieBOT, Citation bot, Xqbot, Anna Frodesiak, Jonesey95, Isiaunia, The Ent, Tcazes, NGPriest, EdoBot, ClueBot NG, Northamerica1000, SFK2, Cjbukows, Laurieds, Harmoniclag, Olenyash, Monkbot, Jbanegas and Anonymous: 14

- **Forest gardening** *Source:* https://en.wikipedia.org/wiki/Forest_gardening?oldid=748057418 *Contributors:* Ray Van De Walker, Anthere, Quercusrobur, Lquilter, Stan Shebs, Artost, Glenn, Marshman, Vardion, Alan Liefting, Everyking, Bobblewik, Serendeva, Pgan002, Mike Rosoft, Chris j wood, Guanabot, Bender235, Eadmund~enwiki, Erauch, Sumalsn, Anthony Appleyard, Velella, Kazvorpal, Bobrayner, Rtdrury, Benjitz, Salix alba, Gaius Cornelius, Dialectric, Nirvana2013, Kevin, SmackBot, Cacuija, Lotusduck, Chris the speller, Bluebot, Brimba, Abrahami, Byelf2007, Dandelion1, SilkTork, Gobonobo, Rkmlai, DabMachine, Lograph, Doug Weller, Marek69, Ingolfson, Daniel Cordoba- Bahle, Sustainableyes, Skier Dude, Madbishop, Jorfer, Woodsguy, Scott Roy Atwood, DASonnenfeld, Lightmouse, Der Golem, Mild Bill Hiccup, XLinkBot, Edibleforests, Addbot, Granitethighs, Jarble, Thiestru, Bermicourt, Luckas-bot, AnomieBOT, Rubinbot, Citation bot, Anna Frodesiak, Legion23, BoundaryRider, Citation bot 1, I dream of horses, Lotje, Vrenator, RjwilmsiBot, John of Reading, Look2See1, EME44, Mmeijeri, Lexandalf, ZéroBot, Popok75, Walter Ralt, ClueBot NG, PaleCloudedWhite, Helpful Pixie Bot, Philospelunk, Wbm1058, Gob Lofa, Lavenderdawn, Northamerica1000, Mr. Joca, Rowan Adams, Dexbot, Sminthopsis84, Lisamd, Johnscotaus, Bleu8, Yackityyack, Ginsuloft, JoeHebda, Julietdeltalima, EChastain, JoannaHoman, Bender the Bot and Anonymous: 57

- **Inga alley cropping** *Source:* https://en.wikipedia.org/wiki/Inga_alley_cropping?oldid=620595708 *Contributors:* Bearcat, Stuartyeates, Bg- white, Nirvana2013, Tony1, Nick Number, Katharineamy, Elisevil, IceUnshattered, Jarble, Yobot, Gongshow, Koyos, Craig Pemberton, Vidimian, Kibi78704, Mmeijeri, AvicAWB, Mastermindsarkar.100, JanGolden, Tom Pippens, Tommy Pinball and Anonymous: 15

- **Intercropping** *Source:* https://en.wikipedia.org/wiki/Intercropping?oldid=752533114 *Contributors:* Dysprosia, Pollinator, Alan Liefting, Ger- men, Erauch, Vortexrealm, Maureen, Kjkolb, Linmhall, Kazvorpal, Jwanders, Strait, Salix alba, Chobot, Bgwhite, Kummi, Gaius Cornelius, Nirvana2013, Lockesdonkey, Garion96, SmackBot, McGeddon, CRKingston, Eskimbot, Cacuija, Gilliam, Chris the speller, Darth Panda, Jhml, KP Botany, Smartse, Steven Walling, Wassupwestcoast, Naohiro19, Mbuckingham, Markisgreen, Mikemoral, Der Golem, Mild Bill Hiccup, Tanketz, EmmaRubu, Kembangraps, Addbot, Tassedethe, BlazerKnight, Dyorkey, Materialscientist, Apothecia, Anna Frodesiak,

Craig Pemberton, Ezhuttukari, Katach, Kibi78704, RjwilmsiBot, Skamecrazy123, Look2See1, Slightsmile, Dcirovic, Anir1uph, Matthewc-girling, Jsayre64, عمرو بن كلثوم, ClueBot NG, Minerv, Wbm1058, Gob Lofa, Northamerica1000, Tom Pippens, Jalaber, SMARTY 123, अनुनाद सिंह, YiFeiBot, Jenjhall, TheEditor867, Weopi, Utters11, Bender the Bot, Carsonac, Zcarstvnz and Anonymous: 28

- **Mycoforestry** *Source:* https://en.wikipedia.org/wiki/Mycoforestry?oldid=750936455 *Contributors:* BlueCanoe, Gaius Cornelius, Nirvana2013, Gobonobo, Cydebot, DASonnenfeld, LilHelpa, Anna Frodesiak, Claysulak, Look2See1, Helpful Pixie Bot, Northamerica1000, ArmbrustBot, Bender the Bot and Anonymous: 5

- **Polyculture** *Source:* https://en.wikipedia.org/wiki/Polyculture?oldid=727960519 *Contributors:* Hyacinth, Alan Liefting, Quadell, Pak21, Er-auch, Cmdrjameson, TheParanoidOne, Kazvorpal, Jwanders, Porphyra, Salix alba, Nirvana2013, Calvin08, CmdrObot, Jhml, Nocompost, R'n'B, Uncle Dick, Chiswick Chap, Squids and Chips, ClueBot, XLinkBot, Addbot, Tassedethe, Tikar aurum, Luckas-bot, Apothecia, Anna Frodesiak, Rickproser, האחד והיחיד, LucienBOT, JobenCitySchlicka, Glacier2009, MarcelB612, LESS Productions, Lopifalko, Look2See1, ClueBot NG, Gob Lofa, Northamerica1000, Sminthopsis84, Redddbaron, Will-o-the-west, Monkbot, Potatop467 and Anonymous: 25

- **Riparian buffer** *Source:* https://en.wikipedia.org/wiki/Riparian_buffer?oldid=749425165 *Contributors:* Skysmith, Tpbradbury, Tom, Rjwilmsi, Vmenkov, Nirvana2013, Shepazu, Epipelagic, SmackBot, Sadads, Werdan7, Levineps, Cydebot, Leolaursen, Lady Mondegreen, DASonnen-feld, Lightmouse, Taroaldo, Moreau1, Stobin2, 5 albert square, Jarble, Martess, Minnecologies, FrescoBot, John of Reading, Look2See1, Dcirovic, ClueBot NG, Wbm1058, MusikAnimal, BattyBot, Xyzspaniel, Khazar2, Sidelight12, CTnative, Jbanegas, Anna Mayerhof, Extem-poralist, Tom Santangelo and Anonymous: 15

- **Silvopasture** *Source:* https://en.wikipedia.org/wiki/Silvopasture?oldid=724042601 *Contributors:* Kku, Zoicon5, MarkGallagher, Nirvana2013, Chendy, GRuban, Vogelsad, Bwpach, Winnowhead, Crowish, Fabrictramp, Djibrilla, Look2See1, 28bot, BG19bot, Northamerica1000, Jbane-gas, Anna Mayerhof and Anonymous: 6

- **Trees 4 Children** *Source:* https://en.wikipedia.org/wiki/Trees_4_Children?oldid=751914639 *Contributors:* BD2412, RFD, DASonnenfeld, Jax 0677, BG19bot, DavidLeighEllis, Llbellavin and Bender the Bot

- **Windbreak** *Source:* https://en.wikipedia.org/wiki/Windbreak?oldid=744289107 *Contributors:* Tedernst, Alan Liefting, MPF, Wouterhagens, H-2-O, PFHLai, Viriditas, Grutness, Cuyaya, Mindmatrix, DialUp, Guanxi, Vmenkov, Wavelength, NawlinWiki, Nirvana2013, FF2010, JoshuaGarton, SmackBot, Delldot, Gilliam, Persian Poet Gal, Archibald Tuttle, Dogears, JForget, Plantsurfer, Spojrzenie, Engineman, Drm310, MartinBot, Jay Litman, CommonsDelinker, Leaflet, Shawn in Montreal, Martial75, Rdfr, VolkovBot, Martin451, Flyer22 Reborn, Forest Ash, ClueBot, Hysocc, Excirial, Moreau1, Editorofthewiki, Addbot, Lightbot, Luckas-bot, Yobot, Minnecologies, LilHelpa, Xqbot, Srich32977, GrouchoBot, A.amitkumar, Haeinous, Facklere, Afishslappedme, WikitanvirBot, Look2See1, Dewritech, Ego White Tray, ChuispastonBot, Aigipan, ClueBot NG, Rezabot, Helpful Pixie Bot, Curb Chain, BG19bot, Wightsails, Rm1271, Gorthian, Hadi Payami, Pratyya Ghosh, Ytic nam, Monkbot, JayDee.UU, Jbanegas, KasparBot, FishStampLover52 and Anonymous: 36

- **World Agroforestry Centre** *Source:* https://en.wikipedia.org/wiki/World_Agroforestry_Centre?oldid=745966069 *Contributors:* TimR, Urhix-idur, GregorB, Nirvana2013, Pegship, SmackBot, Ohconfucius, Dl2000, CmdrObot, Nick Number, Dr. Blofeld, Kleomarlo, T L Miles, Mr random, The Anomebot2, Tikiwont, Grmanners, DASonnenfeld, A4bot, Janus01, Moonriddengirl, MichaelHailu, Dana boomer, Ad-dbot, Lightbot, Luckas-bot, Yobot, Savonneux, Katach, Lotje, Minimac, Mean as custard, EmausBot, ZéroBot, Monsoon Waves, BG19bot, Northamerica1000, World Agroforestry Centre, Paulstapleton, BacLuong, AnneWachira and Anonymous: 4

- **Paul Yeboah** *Source:* https://en.wikipedia.org/wiki/Paul_Yeboah?oldid=746463866 *Contributors:* Magioladitis, Yobot, BG19bot, Celestine-sucess, Adjoajo and Anonymous: 2

19.7.2 Images

- **File:2011-06-07-FF-meeting-124wiki1080.jpg** *Source:* https://upload.wikimedia.org/wikipedia/commons/5/58/2011-06-07-FF-meeting-124wiki1080. jpg *License:* CC BY-SA 3.0 *Contributors:* Beacon Food Forest / Original photo by Bob Redmond / Modified by Daniel Johnson *Original artist:* Friends of Beacon Food Forest

- **File:2011-06-07-FF-meeting-133wiki1080.jpg** *Source:* https://upload.wikimedia.org/wikipedia/commons/0/0c/2011-06-07-FF-meeting-133wiki1080. jpg *License:* CC BY-SA 3.0 *Contributors:* Beacon Food Forest / Original photo by Bob Redmond / Modified by Daniel Johnson *Original artist:* Friends of Beacon Food Forest

- **File:2011-06-07-FF-meeting-37wiki1080.jpg** *Source:* https://upload.wikimedia.org/wikipedia/commons/8/8d/2011-06-07-FF-meeting-37wiki1080. jpg *License:* CC BY-SA 3.0 *Contributors:* Beacon Food Forest / Original photo by Bob Redmond / Modified by Daniel Johnson *Original artist:* Friends of Beacon Food Forest

- **File:2011-06-07-FF-meeting-54wiki1080.jpg** *Source:* https://upload.wikimedia.org/wikipedia/commons/2/27/2011-06-07-FF-meeting-54wiki1080. jpg *License:* CC BY-SA 3.0 *Contributors:* Beacon Food Forest / Original photo by Bob Redmond / Modified by Daniel Johnson *Original artist:* Friends of Beacon Food Forest

- **File:Agrosylviculture_australie_Clive_Wawn.jpg** *Source:* https://upload.wikimedia.org/wikipedia/commons/b/b6/Agrosylviculture_australie_ Clive_Wawn.jpg *License:* GFDL *Contributors:* english wikipedia *Original artist:* Lamiot, with original pictures made by Clive_Wawn

- **File:Alley_cropping_corn_walnuts.jpg** *Source:* https://upload.wikimedia.org/wikipedia/commons/e/e9/Alley_cropping_corn_walnuts.jpg *License:* Public domain *Contributors:* What is Alley Cropping? USDA National Agroforestry Center. February 2012 *Original artist:* USDA NAC

- **File:Amanita_praecox_86186.jpg** *Source:* https://upload.wikimedia.org/wikipedia/commons/d/de/Amanita_praecox_86186.jpg *License:* CC BY-SA 3.0 *Contributors:* This image is Image Number 86186 at Mushroom Observer, a source for mycological images.
  Original artist: This image was created by user Dan Molter (shroomydan) at Mushroom Observer, a source for mycological images.

- **File:Ambox_important.svg** *Source:* https://upload.wikimedia.org/wikipedia/commons/b/b4/Ambox_important.svg *License:* Public domain *Contributors:* Own work, based off of Image:Ambox scales.svg *Original artist:* Dsmurat (talk · contribs)

19.7.3 Content license